Lecture Notes in Mathematics 1544

Michael Schürmann

White Noise on Bialgebras

Springer-Verlag

Editorial Policy

for the publication of monographs

In what follows all references to monographs, are applicable also to multiauthorship volumes such as seminar notes.

§ 1. Lecture Notes aim to report new developments - quickly, informally, and at a high level. Monograph manuscripts should be reasonably self-contained and rounded off. Thus they may, and often will, present not only results of the author but also related work by other people. Furthermore, the manuscripts should provide sufficient motivation, examples and applications. This clearly distinguishes Lecture Notes manuscripts from journal articles which normally are very concise. Articles intended for a journal but too long to be accepted by most journals, usually do not have this "lecture notes" character. For similar reasons it is unusual for Ph. D. theses to be accepted for the Lecture Notes series.

§ 2. Manuscripts or plans for Lecture Notes volumes should be submitted (preferably in duplicate) either to one of the series editors or to Springer- Verlag, Heidelberg . These proposals are then refereed. A final decision concerning publication can only be made on the basis of the complete manuscript, but a preliminary decision can often be based on partial information: a fairly detailed outline describing the planned contents of each chapter, and an indication of the estimated length, a bibliography, and one or two sample chapters - or a first draft of the manuscript. The editors will try to make the preliminary decision as definite as they can on the basis of the available information.

§ 3. Final manuscripts should preferably be in English. They should contain at least 100 pages of scientific text and should include

- a table of contents;
- an informative introduction, perhaps with some historical remarks: it should be accessible to a reader not particularly familiar with the topic treated;
- a subject index: as a rule this is genuinely helpful for the reader.

Further remarks and relevant addresses at the back of this book.

Lecture Notes in Mathematics 1544

Michael Schürmann

White Noise on Bialgebras

Springer-Verlag
Berlin Heidelberg New York
London Paris Tokyo
Hong Kong Barcelona
Budapest

Author

Michael Schürmann
Institut für Angewandte Mathematik
Universität Heidelberg
Im Neuenheimer Feld 294
W-6900 Heidelberg, Germany

Mathematics Subject Classification (1991): 81S25, 81R50, 60J30, 81S05, 60B15

ISBN 3-540-56627-9 Springer-Verlag Berlin Heidelberg New York
ISBN 0-387-56627-9 Springer-Verlag New York Berlin Heidelberg

Printed in Germany

Typesetting: Camera-ready by author/editor
46/3140-543210 - Printed on acid-free paper

To Jutta, Antje and Matthias

CONTENTS

Introduction

These notes are a contribution to the field of quantum (or non-commutative) probability theory. Quantum probability can be regarded as an attempt of a unified approach to classical probability and the quantum theory of irreversible processes.

Our special interest lies in a non-commutative theory of processes with independent and stationary increments on a group. Following common practise, such processes will be called *white noise* even though they are actually 'integrated white noise'. The motivating idea is to use non-commutative white noise as a description of a quantum mechanical heat bath to which the quantum mechanical initial system under consideration is coupled. The passage from classical commutative to quantum non-commutative theory is mathematically established by replacing algebras of functions by algebras of linear operators, i.e. by not necessarily commutative algebras. The natural generalization of an algebra of functions on a semi-group (a group), and thus in a way of a semi-group (a group) itself, is a *bialgebra* (a *Hopf algebra*). Recently, *quantum groups* appeared in several fields of non-commutative mathematics; see e.g. [25, 53, 94]. It should be stressed that quantum groups are defined in different ways in the literature but they are Hopf algebras in all cases. We are concerned with the role bialgebras and Hopf algebras play in quantum probability as the 'non-commutative state space' on which our non-commutative white noise is modelled. Since we work with generalizations of probability measures, there is also a need for a positivity structure, and this is how $*$-bialgebras and $*$-Hopf algebras come in. These are bialgebras and Hopf algebras which have an involution compatible both with the algebra and the coalgebra structure. A $*$-Hopf algebra is close to what is called a *matrix pseudo-group* in [94].

The first time $*$-bialgebras appeared in quantum probability was in the paper [92] by W. von Waldenfels where the *non-commutative coefficient algebra of the unitary group* was introduced. After a crucial result on positivity had been established [91, 70], L. Accardi proposed an algebraic framework for the general theory of quantum white noise on graded $*$-bialgebras. This program has been worked out in [5].

One of the main tools in quantum probability is the *quantum stochastic calculus* developed by R.L. Hudson and K.R. Parthasarathy in [41]; see also [15, 16, 51]. The connection between quantum white noise and quantum stochastic differential equations became apparent in [72]. Shortly after that, the question arose as to whether any white noise on a $*$-bialgebra can be realized as a solution of a quantum stochastic differential equation. Under the assumptions of boundedness and of Bose independence of the processes, an answer in the affirmative was given in [29]. Generalizing the results of [75], in these notes we treat the case of arbitrary unbounded processes with 'twisted' independence. We show that a white noise can always be realized on Bose Fock space. Using H. Maassens kernel method [51], we give an explicit formula for the processes. Indeed, the processes are solutions of quantum stochastic differential equations. The form of the equation is governed by the coalgebra structure of the underlying twisted $*$-bialgebra.

Recently, there has been an attempt to generalize both the concepts of convolution semi-groups of states on a $*$-bialgebra (which are closely related to quantum white noise) and of convolution semi-groups of instruments on a group [11] to a theory of convolution semi-groups of positive operator valued maps on a $*$-bialgebra; see [12].

*

The following discussion of four related topics in classical probability theory on groups and in quantum probability is intended to serve as a more detailed introduction to the mathematical problems we are concerned with.

Convolution semi-groups of probability measures. Let $X_t : \Omega \to G$ be a stochastic process indexed by time $t \geq 0$ taking values in some (topological) group G. Suppose that the process X_t is a white noise so that it has independent and stationary increments

$$X_{st} = X_s^{-1} X_t,$$

i.e. the random variables

$$X_{t_1 t_2}, \ldots, X_{t_n t_{n+1}}$$

are independent for all choices of $n \in \mathbb{N}$ and $t_1 < \cdots < t_{n+1}$ and the distribution of X_{st} only depends on the difference $t - s$. Then X_t is determined up to stochastic equivalence by its 1-dimensional distributions φ_t which form a 1-parameter semi-group of probability measures on G with respect to convolution. If φ_t is weakly continuous at 0 we can differentiate $\varphi_t(f)$ for an appropriate class of complex-valued functions f on G to obtain the *generator* of φ_t which in all cases of interest again will determine φ_t and therefore X_t.

For a Lie group G *Hunt's formula* [44] gives a description of all the generators of white noise in terms of left invariant derivations of first and second order and an integral which represents the 'Poisson part' of X_t.

For a compact or locally compact abelian group the generator can be defined on the algebra $\mathcal{R}(G)$ of representative functions (see [34]) of G and the generators of white noise turn out to be the hermitian, conditionally positive linear functionals on $\mathcal{R}(G)$; cf. [35]. The principle of correspondence between 1-parameter semi-groups of positive linear functionals and hermitian, conditionally positive linear functionals goes back to I. Schoenberg [69] and will be called *Schoenberg correspondence*; cf. [17].

Stochastic semi-groups. Let (X_{st}) be a stochastic process indexed by pairs (s,t) of real numbers with $0 \leq s \leq t$, taking values in the space M_n of complex $n \times n$-matrices. Suppose that X_{st} satisfies the evolution equations

$$X_{rs} X_{st} = X_{rt}$$
$$X_{tt} = 1$$

almost everywhere for all $r \leq s \leq t$ and that the 'increments' X_{st} are independent. A process X_{st} with these properties is usually called a stochastic semi-group; see e.g. [81]. Let the increments X_{st} also be stationary and assume that X_{st} converges to 1 in probability for $t \downarrow 0$. (In this case, X_{st} is actually the increment process associated with a classical white noise on the general linear group.) Then there exists a stochastic process $(F_t)_{t \geq 0}$ on the same probability space taking values in M_n such that the following holds. The additive increments $F_{st} = F_t - F_s$ are independent and stationary, F_t converges to

0 in probability for $t \downarrow 0$ and X_{st} is the solution of the operator stochastic differential equation

$$\mathrm{d}X_{st} = X_{st}\mathrm{d}F_t, \quad t \geq s, \tag{1}$$

with the initial condition $X_{ss} = 1$ a.e. Moreover, the sums

$$\sum_{0 \leq t_{k+1} < t} (X_{t_k, t_{k+1}} - 1)$$

converge to F_t in probability if the maximal distance between neighbours of the partition $\{0 = t_0 < t_1 < t_2 < \ldots\}$ of $\mathbb{R}_+$ tends to 0; see [81] where the more general case is treated when X_{st} are not necessarily stationary but for each fixed s the process $(X_{st})_{s \leq t}$ is a semi-martingale.

Infinitely divisible representations of groups. A normalized positive definite function f on the group G is called infinitely divisible if for all $n \in \mathbb{N}$ there is a normalized positive definite function f_n on G such that

$$f_n(x)^n = f(x) \text{ for all } x \in G.$$

By Schoenberg correspondence $f = \mathrm{e}^g$ is infinitely divisible if g is hermitian, conditionally positive and g vanishes on the identity element of G. Under certain analytic conditions on f or G each infinitely divisible function on G is of the form e^g; see [63]. The question we are interested in is how to express the Gelfand-Naimark-Segal-representation, see [67], given by f in terms of the following representation naturally associated with g. Divide the vector space $\mathbb{C}G$ formally spanned by the elements of G by the nullspace of the positive sesquilinear form on $\mathbb{C}G$ with

$$(x, y)_g = g(x^{-1}y) - \overline{g(x)} - g(y), \quad x, y \in G,$$

to obtain a pre-Hilbert space D whose completion we denote by H. Denote the canonical mapping from $\mathbb{C}G$ to H by η. Then the equations

$$\rho(x)\eta(y) = \eta(xy) - \eta(x)$$

define a unitary representation ρ of G. Roughly speaking, we must *exponentiate* the representation ρ. The so-called *Araki-Woods embedding theorem* for infinitely divisible functions on a group, see [8, 9, 32, 63, 85], says that the GNS-representation of f lives in the *Bose Fock space* over H. Moreover, it maps exponential vectors to multiples of exponential vectors.

Now quantum probability comes in. The quantum stochastic calculus of Hudson and Parthasarathy is a theory of Ito type integration against operator processes on Bose Fock space over $L^2(\mathbb{R}_+) \otimes H$ where H is some Hilbert space fixing the number of degrees of freedom of the integrators. For $\xi \in H$ and a linear operator R on H the integrators are the *creation process* $A_t^*(\xi)$, the *preservation process* $\Lambda_t(R)$, the *annihilation process* $A_t(\xi)$ and time t, the latter giving the usual Riemann-Bochner integral of operators. It is well known that $A_t^* + A_t$ is the realization of Brownian motion on Bose Fock space; see e.g. [36]. Moreover, $A_t^* + \Lambda_t + A_t + t$ is the realization of a Poisson process on Bose Fock space [41]. So the Hudson-Parthasarathy integral generalizes classical stochastic integration against the Wiener and the Poisson process. (We will see that a much wider class of classical types of white noise is included.) The connection to infinitely divisible representations of G is given by the fact that the GNS-representation π_1 of f can be embedded into the solution of the linear quantum stochastic differential equation

$$\mathrm{d}\pi_t(x) = \pi_t(x)\mathrm{d}F_t(x) \tag{2}$$

with initial condition $\pi_0(x) = \mathrm{id}$ where

$$F_t(x) = A_t^*(\eta(x)) + \Lambda_t(\rho(x) - \mathrm{id}) + A_t(\eta(x^{-1})) + g(x)t.$$

Finally this makes rigorous the idea of exponentiating the representation ρ; cf. [59].

Unification of Bose and Fermi stochastic calculus. It is possible to develop a quantum stochastic calculus on Fermi Fock space [6, 7]; see also the work of C. Barnett, R. Streater and I. Wilde [14] on the Clifford integral. However, it was shown in [43] that the Bose does include the Fermi calculus. This unification is possible because Fermion Brownian motion itself can be interpreted as the Bose quantum stochastic integral

$$\int_0^t \Gamma_\tau(-1)\mathrm{d}A_\tau$$

over the second quantization $\Gamma_t(-1)$ of the *reflexion process* on $L^2(\mathbb{R}_+)$ which is given by

$$f \mapsto -f\chi_{[0,t)} + f\chi_{[t,\infty)}.$$

*

Up to this point we only arranged some known facts of classical probability, mentioned a relation to quantum stochastic calculus and quoted the result on the unification of Bose and Fermi integration. We explain how all this can be seen in a more general non-commutative framework and finally becomes a mathematical theory which goes far beyond the classical results.

For motivation and in order to simplify things, for the moment let us restrict ourselves to the case when G is the group U_n of complex unitary $n \times n$-matrices. Let $\mathcal{U}[n]$ be the $*$-algebra of complex-valued functions on U_n generated by the constant functions and by the coordinate functions x_{kl}, $k, l = 1, \ldots, n$, and their complex-conjugates x_{kl}^*. Since U_n is compact, probability measures on U_n and states, i.e. normalized positive linear functionals, on $\mathcal{U}[n]$ can be identified; see e.g. [35]. The convolution of probability measures becomes the convolution of states φ and ψ given by

$$\varphi \star \psi = (\varphi \otimes \psi) \circ \Delta$$

where Δ denotes the $*$-algebra homomorphism

$$\Delta : \mathcal{U}[n] \to \mathcal{U}[n] \otimes \mathcal{U}[n]$$

given by extension of

$$\Delta x_{kl} = \sum_{m=1}^{n} x_{km} \otimes x_{ml}. \tag{3}$$

The generators of white noise on U_n are the hermitian linear functionals ψ on $\mathcal{U}[n]$ with the property

$$\psi(b^* b) \geq 0 \text{ for all } b \in \text{Kern } \epsilon$$

where ϵ denotes the $*$-algebra homomorphism

$$\epsilon : \mathcal{U}[n] \rightarrow \mathbb{C}$$

given by extension of

$$\epsilon(x_{kl}) = \delta_{kl}. \tag{4}$$

The equation

$$(X(\omega))_{kl} = U_{kl}(\omega)$$

establishes a 1-1-correspondence between U_n-valued random variables X and unitary elements U of the $*$-algebra $\mathrm{M}_n \otimes L^\infty$ formed by the $n \times n$-matrices with entries in the commutative von Neumann algebra consisting of the bounded measurable functions on the underlying probability space. The non-commutative analogue of a process $(U_t)_{t\geq 0}$ with $U_t \in \mathrm{M}_n \otimes L^\infty$ is a family $(U_t)_{t\geq 0}$ of unitary elements in a von Neumann algebra of the form $\mathrm{M}_n \otimes \mathcal{N}$ where now $\mathcal{N}$ is an arbitrary, not necessarily commutative von Neumann algebra. In other words, the non-commutative generalization of a stochastic process on U_n is a family $(U_t)_{t\geq 0}$ of unitary operators on a Hilbert space of the form $\mathbb{C}^n \otimes \mathcal{H}$ with $\mathcal{H}$ a Hilbert space. The role of the probability measure is then played by a unit vector Ω in $\mathcal{H}$; in the commutative case $\mathcal{H}$ is the space of square integrable functions and $\Omega = 1$.

It is now important that $\mathcal{U}[n]$ is isomorphic to the (complex, associative, unital) algebra generated by the commuting indeterminates x_{kl} and x^*_{kl} satisfying the unitarity relations

$$\begin{aligned}\sum_{m=1}^{n} x_{km}x^*_{lm} &= \delta_{kl}\mathbf{1}\\ \sum_{m=1}^{n} x^*_{mk}x_{ml} &= \delta_{kl}\mathbf{1};\end{aligned} \tag{5}$$

see e.g. [30]. The non-commutative analogue of $\mathcal{U}[n]$ is the algebra $\mathcal{U}\langle n\rangle$ generated by *non-commuting* indeterminates x_{kl} and x^*_{kl} with relations (5). Then $\mathcal{U}\langle n\rangle$ is isomorphic to the algebra generated by the coordinate functions

$$x_{kl} : U(\mathbb{C}^n \otimes \mathcal{H}) \rightarrow \mathbf{B}(\mathcal{H})$$

and their adjoints x^*_{kl} given by $x^*_{kl}(U) = (U_{kl})^*$ where $\mathcal{H}$ is an infinite-dimensional Hilbert space, $U(\mathbb{C}^n \otimes \mathcal{H})$ denotes the group of unitary operators on $\mathbb{C}^n \otimes \mathcal{H}$ and the multiplication in $\mathcal{U}\langle n\rangle$ is given by pointwise multiplication in the algebra $\mathbf{B}(\mathcal{H})$ of bounded linear operators on $\mathcal{H}$; see [30].

Again we can define $*$-algebra homomorphisms Δ and ϵ by extension of (3) and (4). Both $\mathcal{U}[n]$ and $\mathcal{U}\langle n\rangle$ are examples of $*$-bialgebras. The general notion of a $*$-bialgebra comprises a $*$-algebra $\mathcal{B}$ and $*$-algebra homomorphisms

$$\Delta : \mathcal{B} \rightarrow \mathcal{B} \otimes \mathcal{B}$$

and

$$\epsilon : \mathcal{B} \to \mathbb{C}$$

satisfying the *coassociativity* and *counit* conditions; see [1, 87]. The mapping Δ is called the *comultiplication* and ϵ is called the *counit* of $\mathcal{B}$.

A quantum random variable in the sense of [4] on a $*$-algebra $\mathcal{B}$ is a unit preserving $*$-algebra homomorphism from $\mathcal{B}$ to another $*$-algebra $\mathcal{A}$ where $\mathcal{A}$ is also equipped with a state Φ. We can identify elements in $U(\mathbb{C}^n \otimes \mathcal{H})$ and $*$-algebra homomorphisms

$$j_U : \mathcal{U}\langle n\rangle \to \mathbf{B}(\mathcal{H})$$

via the equation

$$j_U(x_{kl}) = U_{kl}.$$

Suppose that Ω is a unit vector in $\mathcal{H}$ giving rise to a state on the $*$-algebra $\mathbf{B}(\mathcal{H})$. Then j_U is a quantum random variable on $\mathcal{U}\langle n\rangle$ and the distribution of U is the state φ_U on $\mathcal{U}\langle n\rangle$ given by

$$\varphi_U(b) = \langle \Omega, j_U(b)\Omega\rangle.$$

$b \in \mathcal{U}\langle n\rangle$. If Ω is cyclic for the von Neumann algebra generated by the entries U_{kl} of U then U is determined by φ_U.

Next a non-commutative notion of independence is needed. For the moment we restrict ourselves to Bose independence; cf. [23]. The sub-$*$-algebras $\mathcal{A}_1, \ldots, \mathcal{A}_m$ of a $*$-algebra $\mathcal{A}$ are said to be independent (in the state Φ on $\mathcal{A}$) if

$$\Phi(a_1 \ldots a_m) = \Phi(a_1) \ldots \Phi(a_m)$$

for $a_1, \ldots, a_m \in \mathcal{A}$, and if

$$a_k a_l = a_l a_k \tag{6}$$

for $k \neq l$. Notice that (6) does not mean reduction to the commutative case, since the algebras $\mathcal{A}_k$ themselves are allowed to be non-commutative.

Now the non-commutative generalization of classical unitary white noise is a family $(U_t)_{t\geq 0}$ of unitary operators on $\mathbb{C}^n \otimes \mathcal{H}$ such that the ranges of $j_{t_1 t_2}, \ldots, j_{t_n t_{n+1}}$ are independent for all $n \in \mathbb{N}$, $t_1 < \cdots < t_{n+1}$, the distribution of U_{st} only depends on $t - s$, and φ_t converges weakly, i.e. pointwise, to the distribution of $\mathbf{1}$ for $t \downarrow 0$ where we put $j_{st} = j_{U_{st}}$ and $\varphi_t = \varphi_{U_t}$.

This notion of white noise can easily be formulated for arbitrary $*$-bialgebras $\mathcal{B}$. A white noise on a $*$-bialgebra is a family of $(j_{st})_{0\geq s\geq t}$ of quantum random variables on $\mathcal{B}$ with the same properties of independence, stationarity and weak continuity as before; the generalization of the evolution property is the *increment property*

$$j_{rs} \star j_{st} = j_{rt} \text{ for } r \leq s \leq t$$
$$j_{ss} = \epsilon \mathbf{1},$$

where the $\star$ is the usual convolution product of linear mappings from a coalgebra to an algebra; see [87]. As in the classical case, the white noise is determined, up to quantum stochastic equivalence, by its generator which is a hermitian, conditionally positive linear functional on $\mathcal{B}$. Here conditionally positive means positive on the kernel of the counit ϵ of $\mathcal{B}$.

Clearly, the algebra $\mathbb{C}G$ also fits into this framework. It is a $*$-bialgebra with Δ and ϵ given by $\Delta x = x \otimes x$ and $\epsilon x = 1$ for $x \in G$ respectively. Moreover, the notion of conditionally positive functions on G is a special case of our more general definition for arbitrary $*$-bialgebras. Notice also that $\mathbb{C}G$ is cocommutative, i.e. the range of the comultiplication is included in the symmetric tensor product.

Now we come to a very important point. The construction of the pre-Hilbert space D, the canonical mapping η and the representation ρ out of a conditionally positive linear functional (which we described above for the case of $\mathbb{C}G$) goes through for arbitrary $*$-bialgebras. Indeed, it works for any $*$-algebra and any $*$-algebra homomorphism ϵ from the algebra to the underlying field; the comultiplication is not at all needed for this construction.

There is another important class of cocommutative $*$-bialgebras which will give the notion of quantum additive white noise and which is actually needed if we want to formulate the analogue of the classical result on stochastic semi-groups. It could be called the infinitesimal form of the group $*$-bialgebra $\mathbb{C}G$. Let V be a complex vector space and assume that there is also given a real structure on V which is expressed by the presence of an involution on V, i.e. an antilinear selfinverse mapping $v \mapsto v^*$. We can form the universal object called the tensor $*$-algebra $\mathcal{T}(V)$ over V (see [18, 74]) and turn $\mathcal{T}(V)$ into a cocommutative $*$-bialgebra by extending $\Delta v = v \otimes \mathbf{1} + \mathbf{1} \otimes v$ and $\epsilon v = 0$, $v \in V$, to $*$-algebra homomorphisms.

We explain why tensor algebras can serve as the bialgebras for additive white noise. Suppose $X = (X^{(1)}, \ldots, X^{(n)})$ is an $\mathbb{R}^n$-valued random variable. Then the components $X^{(1)}, \ldots, X^{(n)}$ give rise to self-adjoint operators $F^{(1)}, \ldots, F^{(n)}$ on L^2 which equal multiplication by $X^{(1)}, \ldots, X^{(n)}$ respectively and which commute in the sense that their spectral projections commute. Via the spectral theorem the joint distributions of $X^{(1)}, \ldots, X^{(n)}$ can be recovered from the operators $F^{(1)}, \ldots, F^{(n)}$. This relies on the fact that $F^{(1)}, \ldots, F^{(n)}$ are multiplication operators. It is not clear what a non-commutative generalization of this should be, if we want joint distribution in the sense of [4] to exist. A way out of these difficulties is possible if we restrict our interest to *moments*. Let us assume that X_t is an additve white noise on $\mathbb{R}^n$ such that all the moments

$$m(t; l_1, \ldots, l_n) = \int_\Omega (X_t^{(1)})^{l_1} \ldots (X_t^{(n)})^{l_n} \mathrm{d}P$$

$t \geq 0$, $l_1, \ldots, l_n \in \mathbb{N}_0 = \mathbb{N} \cup \{0\}$, of the process exist. We form the $*$-vector space $V = \mathbb{C}^n$ with the involution given by componentwise complex conjugation. Then $\mathcal{T}(V)$ is the $*$-algebra of polynomials in n indeterminates $x_1, \ldots, x_n$ with complex coefficients and with the involution given by $x_k^* = x_k$. We define the quantum random variable $j_{st}(x_l)$ on $\mathcal{T}(V)$ by extension of $j_{st}(x_l) = F_{st}^{(l)}$ where $F_{st}^{(l)}$ denotes multiplication by $X_t^{(l)} - X_s^{(l)}$ on the linear subspace of L^2 formed by the algebra of functions generated by all the $X_t^{(l)}$. This gives a white noise on the $*$-bialgebra $\mathcal{T}(V)$ with 1-dimensional distributions φ_t given by

$$\varphi_t((x_1)^{l_1} \ldots (x_n)^{l_n}) = m(t; l_1, \ldots, l_n).$$

This white noise is 'classical' because it vanishes on the ideal generated by the elements $x_k x_l - x_l x_k$. A white noise on a tensor $*$-bialgebra $\mathcal{T}(V)$ is called a non-commutative additive white noise on V.

One of the main results of these notes says that a white noise on an arbitrary $*$-bialgebra $\mathcal{B}$ can be realized as the solution of the quantum stochastic differential equation

$$\mathrm{d}j_{st} = j_{st} \star \mathrm{d}F_t \tag{7}$$

with the initial condition $j_{ss} = \epsilon$. Here F_t is the additive white noise on the $*$-vector space $\mathcal{B}$ with

$$F_t = A_t^* \circ \eta + \Lambda_t \circ (\rho - \epsilon) + A_t \circ \tilde{\eta} + t\psi$$

where $\tilde{\eta}(b) = \eta(b^*)$. Moreover, the correlations of F_t are approximated by the correlations of the sums

$$\sum_{0 \le t_{k+1} < t} (j_{t_k, t_{k+1}} - \epsilon). \tag{8}$$

This is a non-commutative analogue of the result for stochastic semi-groups. Moreover, equation (2) for infinitely divisible group representations is nothing but the special case of (7) when $\mathcal{B}$ is the group algebra $\mathbb{C}G$.

Applying our result to unitary white noise, we obtain the analogue (and generalization) of the classical result for white noise on U_n, namely that U_t is the solution of

$$\mathrm{d}U_t = U_t \mathrm{d}F_t$$

where F_t is an additive white noise on $\mathcal{U}\langle n \rangle$ arranged as an $n \times n$-matrix. Indeed, the latter equation turns out to be exactly of the type considered in [41] for one degree of freedom and in [42] for an infinite number of degrees of freedom, so that the unitary white noises are exactly the unitary cocycles appearing as solutions of quantum stochastic differential equations.

Under the hypothesis that U_t is already given on Bose Fock space, these cocycles have been characterized in [40]. For a finite-dimensional initial space our result is more general, because we do not assume $\mathcal{H}$ to be a Bose Fock space from the beginning, the Fock space structure coming from the properties of U_t itself.

As another application we have the characterization of additive white noise. It turns out to be equal to sums of creation, preservation, annihilation and scalar processes. Thus, under the assumption of Bose independence, Bose quantum stochastic calculus covers all possible theories of non-commutative stochastic integration against white noise. Infinitely divisible representation on Lie algebras (cf. [86]) can be embedded into additive white noise on the universal enveloping Hopf algebra, and our result also yields a characterization of these representations.

In view of classical stochastic integration theory, one may also look for a characterization of quantum martingales. The quantum martingale representation theorems of [39, 64, 65], however, start from processes defined on Fock space, whereas in our representation theorem, as already said, one of the main points is that Fock space structure results from properties of the processes. The 'quantum Lévy characterization' of quantum Brownian motion of [3, 60] goes more into the direction of our result. In [3, 60] the condition of independence of increments is partly replaced by a weaker martingale condition. On the other hand, the class of processes to be characterized is much smaller, and this is why there is an additional condition on the processes replacing P. Lévy's condition of continuity of the trajectories.

In the case of cocommutative $*$-bialgebras (which include the examples $\mathbb{C}G$ and $\mathcal{T}(V)$ but not $\mathcal{U}\langle n \rangle$) the theory simplifies considerably. Cocommutative white noise can always be written as a convolution product of the basic processes on Fock space.

As far as a quantum analogue of Hunt's formula on Lie groups is concerned, the investigations still are in an early state. (The results of [13] go into another direction; not the group is replaced by a non-commutative analogue but scalar valued probability measures are generalized to operator valued measures.) All we can say is that, under certain cohomological conditions, a 'maximal quadratic component' corresponding to the Gauss second order derivation term can be extracted [73]. On the other hand, Hunt's formula and the operator stochastic differential equation (1) satisfied by classical matrix valued white noise are closely related; cf. the Ito decomposition formula in [37] for white noise on a Lie group which was inspired by [72].

Without further mention, in all we said up to this point the algebra structure of the tensor product $\mathcal{B} \otimes \mathcal{B}$ was understood to be the usual one given by

$$(b \otimes c)(b' \otimes c') = bb' \otimes cc'.$$

Notice that here the 'symmetric flip operator'

$$\tau : \mathcal{B} \otimes \mathcal{B} \to \mathcal{B} \otimes \mathcal{B}$$

with

$$\tau(b \otimes c) = c \otimes b$$

plays a role. The same symmetric structure appears in condition (6) which is part of the notion of Bose independence. A first step towards a generalization of this leads to $\mathbb{Z}_2$-graded algebras (or super algebras) and graded tensor products of algebras. In this case we have commutation and anti-commutation relations. In particular, Fermi independence [38] is covered by such a theory.

Further considerations lead to a *semi-direct product* $\mathcal{B} \otimes_{(\gamma,\alpha)} \mathcal{B}$ which can be defined for what we call double-module algebras. The latter are $\mathbb{L}$-graded algebras such that there is also given a representation α of $\mathbb{L}$ in the group of automorphisms of $\mathcal{B}$ compatible with the grading γ. We have 'twisted $*$-bialgebras' and 'twisted white noise'; see also [54]. The *ordered commutation relations* are of the form

$$j_k(b) j_l(c) = j_l(\alpha(\mathrm{d}_b)c) j_k(b) \text{ for } k < l$$

where d_b denotes the degree of the homogeneous element b.

There is a phenomenon for general twisted white noise which does not appear in the Bose or Fermi case. One has to distinguish between the 'left' and the 'right' case. Left white noise has the advantage that 'pyramidally' time ordered correlations factorize which, for physical reasons, is a natural requirement; cf. [68]. Given a right white noise we can associate with it a left white noise on the opposite bialgebra, this left white noise being the time reversion of the original right white noise.

Now the trick is to reduce the twisted case to the symmetric case with the help of the *symmetrization* of $\mathcal{B}$ which as an algebra is the (left or right) semi-direct product of $\mathbb{L}$ with $\mathcal{B}$; cf. the notion of 'bosonization' in [55]. Twisted white noise again can be realized on Bose Fock space as the solution of a quantum stochastic differential equation of the form

$$\mathrm{d}j_{st} = j_{st} \star \mathrm{d}F_t$$

where F_t now is a twisted additive white noise. The additive noise itself is a Bose quantum stochastic integral over a second quantization process given by the action of $\mathbb{L}$. This generalizes the unification of Bose and Fermi quantum stochastic calculus.

A special case are the twisted $*$-bialgebras coming from commutation relations of the form

$$j_k(b)j_l(c) = q\, j_l(c)j_k(b)$$

with q a scalar depending on b and c *and* on the order of k and l. A remarkable example of additive q-white noise is closely related to the Azéma martingales (see [10, 26]) which K.R. Parthasarathy in [61] showed to be solutions of quantum stochastic differential equations. Using the theory of white noise on bialgebras, we prove a 'Donsker invariance principle' for Azéma noise where it should be stressed that convergence is in the sense of convergence of correlations.

We mention that other notions of non-commutative independence than the ones used in these notes have been considered; see [47, 68, 82]. In particular, for the case of 'free' independence in the sense of D. Voiculescu [88] a stochastic integration theory has been developed by B. Kümmerer and R. Speicher [83, 48]. In [31] it was shown that an additive free white noise can always be realized on the full, i.e. not symmetrized or anti-symmetrized, Fock space (but cf. [66] where free white noise is realized on Bose Fock space as a 'non-adapted' quantum stochastic integral). Using the concept of Voiculescu's dual groups (see [89]) rather than bialgebras, a theory similar to the one developed in the present notes seems to be possible for the case of free independence.

These notes are organized as follows. Chapter 1 presents notations, definitions and examples and gives the elementary basic results on non-commutative white noise. Using H. Maasens quantum stochastic kernel calculus and the fundamental theorem on coalgebras, Chapter 2 deals with the realization of symmetric white noise on Bose Fock space. The 'symmetrization trick' can be found in Chapter 3. Chapter 4 contains the application of the results of Chapters 2 and 3 and basic examples such as unitary and additive white noise. In particular, the characterization of the Azéma type noise, of twisted unitary Hudson-Parthasarathy cocycles and of cocommutative white noise is treated in detail. Chapter 5 is on maximal quadratic components and presents an application to infinitely divisible representations of the canonical commutation and anti-commutation relations. Finally, Chapter 6 contains a limit theorem for coalgebras with applications to quantum probability of which the approximation of the underlying additive noise by sums of the form (8) is one. The other application is to prove various functional limit theorems for Azéma martingales and, more generally, for Azéma noise and related types of quantum stochastic white noise.

The author is grateful to all mathematicians who, with their interest and encouragement, helped to make these notes possible. Many ideas grew out of the exchange with other quantum probabilists and with research workers in related fields. Special thanks are due to L. Accardi, D. Applebaum, R.L. Hudson, P.D.F. Ion, S. Majid, K.R. Parthasarathy, J. Quaegebeur, R. Speicher and W. von Waldenfels.

1. Basic concepts and first results.

We begin with some algebraic notions which will be used throughout these notes. Next we introduce the language of quantum probability. Additive and multiplicative white noise are presented as the basic examples of non-commutative white noise. We make the attempt to derive the axioms of non-commutative independence, $*$-bialgebras and white noise from a few 'first assumptions'. A number of examples underlines the richness of the theory. Finally, some basic results such as the existence of a generator and the algebraic reconstruction of a process are proved.

1.1. Preliminaries. Vector spaces will be over the complex numbers. An algebra is a triplet $(\mathcal{A}, M, \kappa)$ consisting of a vector space $\mathcal{A}$ and linear mappings

$$M : \mathcal{A} \otimes \mathcal{A} \to \mathcal{A}$$
$$\kappa : \mathbb{C} \to \mathcal{A}$$

satisfying

$$M \circ (M \otimes \mathrm{id}) = M \circ (\mathrm{id} \otimes M) \qquad \text{(associativity)}$$

and

$$M \circ (\kappa \otimes \mathrm{id}) = M \circ (\mathrm{id} \otimes \kappa) = \mathrm{id}. \qquad \text{(unit property)}$$

One writes $M(a \otimes b) = ab$ and $\kappa(1) = 1$. A coalgebra is a triplet $(\mathcal{C}, \Delta, \epsilon)$ consisting of a vector space $\mathcal{C}$ and linear mappings

$$\Delta : \mathcal{C} \to \mathcal{C} \otimes \mathcal{C}$$
$$\epsilon : \mathcal{C} \to \mathbb{C}$$

satisfying

$$(\Delta \otimes \mathrm{id}) \circ \Delta = (\mathrm{id} \otimes \Delta) \circ \Delta \qquad \text{(coassociativity)}$$

and

$$(\varepsilon \otimes \mathrm{id}) \circ \Delta = (\mathrm{id} \otimes \varepsilon) \circ \Delta = \mathrm{id}. \qquad \text{(counit property)}$$

The mapping Δ is called the *comultiplication* and ϵ is called the *counit* of $\mathcal{C}$. If

$$\Delta b = \sum_{i=1}^{n} b_{1i} \otimes b_{2i}$$

we also write

$$\Delta b = \sum b_{(1)} \otimes b_{(2)}.$$

We put $\Delta_0 = \epsilon$ and define $\Delta_n : \mathcal{C} \to \mathcal{C}^{\otimes n}$, $n \in \mathbb{N}$, inductively by

$$\Delta_{n+1} = (\mathrm{id} \otimes \Delta) \circ \Delta_n.$$

We use the notation

$$\Delta_n b = \sum b_{1i} \otimes \ldots \otimes b_{ni} = \sum b_{(1)} \otimes \ldots \otimes b_{(n)};$$

see [1, 87].

For an algebra $\mathcal{A}$ a vector space V is called a left $\mathcal{A}$-module if there is a linear mapping

$$\alpha : \mathcal{A} \otimes V \to V$$

satisfying

$$\alpha \circ (M \otimes \mathrm{id}) = \alpha \circ (\mathrm{id} \otimes \alpha)$$
$$\alpha \circ (\kappa \otimes \mathrm{id}) = \mathrm{id}$$

or, in other words,

$$\alpha_{a_1 a_2}(v) = \alpha_{a_1}(\alpha_{a_2}(v))$$
$$\alpha_1(v) = v$$

where we used the notation $\alpha(a \otimes v) = \alpha_a(v)$. We also write $a.v$ for $\alpha_a(v)$. Right $\mathcal{A}$-modules are defined similarly.

The symbol $\mathbb{L}$ denotes a group. The identity element of $\mathbb{L}$ is e. By $\mathbb{C}\mathbb{L}$ we denote the algebra consisting of formal finite linear combinations of elements of $\mathbb{L}$ with the multiplication being the extension of the multiplication of $\mathbb{L}$. Then e is the unit element of $\mathbb{C}\mathbb{L}$. A left $\mathbb{C}\mathbb{L}$-module vector space is called a *module vector space.* If we set

$$v.\lambda = \lambda^{-1}.v,$$

$v \in V$, $\lambda \in \mathbb{L}$, we obtain a right $\mathbb{C}\mathbb{L}$-module structure of V.

For a coalgebra $\mathcal{C}$ a vector space V is called a left $\mathcal{C}$-comodule if there is a linear mapping

$$\gamma : V \to \mathcal{C} \otimes V$$

satisfying

$$(\Delta \otimes \mathrm{id}) \circ \gamma = (\mathrm{id} \otimes \gamma) \circ \gamma \tag{1.1.1}$$
$$(\epsilon \otimes \mathrm{id}) \circ \gamma = \mathrm{id}. \tag{1.1.2}$$

The vector space $\mathbb{C}\mathbb{L}$ becomes a coalgebra if we set

$$\Delta\lambda = \lambda \otimes \lambda$$
$$\epsilon\lambda = 1$$

for $\lambda \in \mathbb{L}$ and extend by linearity. A left $\mathbb{C}\mathbb{L}$-comodule vector space is called a *comodule vector space* or *graded vector space.* If we put

$$V^{(\lambda)} = \{v \in V : \gamma(v) = \lambda \otimes v\}$$

we have

$$V = \bigoplus_{\lambda \in \mathbb{L}} V^{(\lambda)}. \tag{1.1.3}$$

Indeed, if we write

$$\gamma(b) = \bigoplus_{\lambda \in \mathbb{L}} \lambda \otimes b_\lambda$$

we have from (1.1.1) and (1.1.2) that $\gamma(b_\lambda) = \lambda \otimes b_\lambda$ and $b = \bigoplus_{\lambda \in \mathbb{L}} b_\lambda$. On the other hand, given a direct sum decomposition (1.1.3) of V we can define a comodule structure of V by $\gamma(b) = \sum_{\lambda \in \mathbb{L}} \lambda \otimes b_\lambda$. An element $v \neq 0$ of $\bigcup_{\lambda \in \mathbb{L}} V^{(\lambda)}$ is called *homogeneous* and for homogeneous v the degree $\mathrm{d}(v) = \mathrm{d}_v$ of v is λ if $v \in V^{(\lambda)}$. The elements of $V^{(e)}$ are called *even*. A vector space can always be equipped with the trivial graduation given by

$$V^{(\lambda)} = \{0\} \text{ unless } \lambda = e \text{ and } V^{(e)} = V.$$

The field $\mathbb{C}$ of complex numbers is always considered as a graded vector space with the trivial graduation. A linear subspace W of V is said to be graded if $W = \bigoplus_{\lambda \in \mathbb{L}} (W \cap V^{(\lambda)})$. Let V and W be two graded vector spaces. The direct sum is again $\mathbb{L}$-graded if we put $(V \oplus W)^{(\lambda)} = V^{(\lambda)} \oplus W^{(\lambda)}$. The vector space $\mathbf{A}(V, W)$ formed by the additive mappings from V to W is $\mathbb{L}$-graded with

$$\mathbf{A}(V, W)^{(\lambda)} = \{R \in \mathbf{A}(V, W) : RV^{(\mu)} \subset W^{(\lambda\mu)} \text{ for all } \mu \in \mathbb{L}\}.$$

We denote by $\mathbf{L}(V, W)$ the $\mathbb{L}$-graded linear subspace of $\mathbf{A}(V, W)$ formed by the linear mappings. A $\mathbb{L}$-graduation of the tensor product $V \otimes W$ is given by

$$(V \otimes W)^{(\lambda)} = \bigoplus_{\mu_1 \mu_2 = \lambda} V^{(\mu_1)} \otimes V^{(\mu_2)}.$$

A vector space is said to be a *double-module* vector space if it is a module and a comodule vector space such that

$$\mathrm{d}(\lambda.v) = \lambda \mathrm{d}(v) \lambda^{-1}.$$

This compatibility condition can also be written

$$\gamma \circ \alpha = (\alpha_{\mathbb{L}} \otimes \alpha) \circ (\mathrm{id} \otimes \tau \otimes \mathrm{id}) \circ (\Delta_{\mathbb{L}} \otimes \gamma)$$

where $\alpha_{\mathbb{L}}$ denotes the module structure of $\mathbb{C}\mathbb{L}$ given by conjugation, i.e.

$$\alpha_{\mathbb{L}}(\lambda_1)\lambda_2 = \lambda_1 \lambda_2 \lambda_1^{-1},$$

the symbol τ stands for the 'flip' operator on $\mathbb{C}\mathbb{L} \otimes \mathbb{C}\mathbb{L}$ defined as the linear operator with $\tau(\lambda_1 \otimes \lambda_2) = \lambda_2 \otimes \lambda_1$, and $\Delta_{\mathbb{L}}$ is the comultiplication of $\mathbb{C}\mathbb{L}$. We mention that the concept of a double-module vector space makes sense if $\mathbb{C}\mathbb{L}$ is replaced by an arbitrary Hopf algebra.

A *module algebra* is an algebra $\mathcal{A}$ which is also a module vector space in such a way that

$$\lambda.(ab) = (\lambda.a)(\lambda.b)$$
$$\lambda.\mathbf{1} = \mathbf{1}$$

for all $\lambda \in \mathbb{L}$ and $a, b \in \mathcal{A}$. In other words, α is a homomorphism from $\mathbb{L}$ to the automorphism group of the algebra $\mathcal{A}$. A *comodule* or *graded algebra* is an algebra which is also a graded vector space such that M and κ are even, i.e.

$$\mathrm{d}(ab) = \mathrm{d}(a)\mathrm{d}(b)$$
$$\mathrm{d}(1) = e.$$

A *module coalgebra* is a coalgebra $\mathcal{C}$ which is also a module vector space in such a way that

$$\Delta(\lambda.c) = \sum \lambda.c_{(1)} \otimes \lambda.c_{(2)}$$
$$\epsilon(\lambda.c) = \epsilon(c)$$

for $\lambda \in \mathbb{L}$ and $c \in \mathcal{C}$. A *comodule* or *graded coalgebra* is a coalgebra $\mathcal{C}$ which is also a graded vector space such that Δ and ε are even. Similarly, there are double-module algebras and double-module coalgebras.

The group $\mathbb{L}$ is called *involutive* or a *$*$-group* if it has an involution, i.e. a mapping $\gamma \mapsto \gamma^*$ with the properties $(\gamma^*)^* = \gamma$ and $(\gamma_1\gamma_2)^* = \gamma_2^*\gamma_1^*$. A group G always can be turned into an involutive group by setting $g^* = g^{-1}$. We denote this involutive group by G^{inv}. If G is an abelian group we always can put $g^* = g$ to make an involutive group out of G which we denote by G^{id}. Finally, for an abelian group G we can turn $G \times G$ into the involutive group $(G \times G)^{\mathrm{twist}}$ by setting $(g, h)^* = (h, g)$. An *involutive* or *$*$-vector space* is a vector space V with an involution, i.e. an anti-linear mapping $v \mapsto v^*$ on V satisfying $(v^*)^* = v$. Let $\mathbb{L}$ be an involutive group. A graded $*$-vector space V is a $*$-vector space V which also is a graded vector space and such that $v \in V^{(\lambda)}$ implies $v^* \in V^{(\lambda^*)}$. For a homogeneous element v the latter condition reads $\mathrm{d}(v^*) = \mathrm{d}(v)^*$. A module $*$-vector space is a module vector space and a $*$-vector space such that

$$(\lambda.v)^* = v^*.\lambda^* = (\lambda^*)^{-1}.v^*.$$

A $*$-algebra is an algebra $\mathcal{A}$ which is also a $*$-vector space such that $(ab)^* = b^*a^*$ for all $a, b \in \mathcal{A}$. A graded $*$-algebra is a $*$-algebra which is also a graded algebra and a graded $*$-vector space. It is clear by now what a module $*$-algebra and a double-module $*$-algebra are.

1.2. Quantum probabilistic notions. A *quantum probability space* is a pair $(\mathcal{A}, \Phi)$ consisting of a $*$-algebra $\mathcal{A}$ and a state (i.e. a normalized positive linear functional) Φ on $\mathcal{A}$.

Let $(\Omega, \mathcal{F}, P)$ be a classical probability space. A $*$-algebra $\mathcal{A}$ of complex-valued functions on Ω which is a linear subspace of $L^1(\Omega, \mathcal{F}, P)$ is a quantum probability space if we take Φ to be the restriction Φ_P to $\mathcal{A}$ of the linear functional $f \mapsto \int_\Omega f \mathrm{d}P$ on $L^1(\Omega, \mathcal{F}, P)$. One possible choice is $\mathcal{A} = L^\infty(\Omega, \mathcal{F}, P)$. Motivated by this example, a quantum probability space sometimes is defined to be a pair $(\mathcal{N}, \Phi)$ consisting of a von Neumann algebra $\mathcal{N}$ and a normal state Φ on $\mathcal{N}$; cf. [4, 47, 52]. This has the advantage that commutative quantum probability spaces, i.e. spaces with commutative $\mathcal{N}$, reduce to spaces of the form $(L^\infty(\Omega, \mathcal{F}, P), \Phi_P)$ that is to an algebra $\mathcal{F}$ of events and a classical probability measure P. On the other hand, we will work with algebras of unbounded operators, for example, if we use the non-commutative moment method. Thus a restriction to von Neumann algebras, for our purposes, seems not to be appropriate in all cases.

Let $\mathcal{D}$ be a pre-Hilbert space. A linear operator F on $\mathcal{D}$ is called *hermitian* if

$$\langle \xi, F\zeta \rangle = \langle F\xi, \zeta \rangle \text{ for all } \xi, \zeta \in \mathcal{D}.$$

Denote by $\mathbf{H}(\mathcal{D})$ the linear subspace of the algebra $\mathbf{L}(\mathcal{D})$ of linear operators on $\mathcal{D}$ spanned by the hermitian linear operators. The space $\mathbf{H}(\mathcal{D})$ is a subalgebra of $\mathbf{L}(\mathcal{D})$, and it consists of all $F \in \mathbf{L}(\mathcal{D})$ with the property that there exists a $G \in \mathbf{L}(\mathcal{D})$ such that

$$\langle \xi, F\zeta \rangle = \langle G\xi, \zeta \rangle \text{ for all } \xi, \zeta \in \mathcal{D}.$$

We set $G = F^*$ and $\mathbf{H}(\mathcal{D})$ becomes a $*$-algebra with involution $F \mapsto F^*$. A representation of a $*$-algebra $\mathcal{A}$ on a pre-Hilbert space $\mathcal{D}$ is a $*$-algebra homomorphism from $\mathcal{A}$ to $\mathbf{H}(\mathcal{D})$; cf. [67]. Applying the Gelfand-Naimark-Segal (GNS-)construction to the pair $(\mathcal{A}, \Phi)$ we obtain a pre-Hilbert space $\mathcal{D}$, a $*$-representation π of $\mathcal{A}$ on $\mathcal{D}$, and a unit vector Ω_0 in $\mathcal{D}$ which is cyclic for $\{\pi(a) : a \in \mathcal{A}\}$ such that

$$\Phi(a) = \langle \Omega_0, \pi(a)\Omega_0 \rangle \text{ for all } a \in \mathcal{A}.$$

A *quantum random variable* over a quantum probability space $(\mathcal{A}, \Phi)$ on a $*$-algebra $\mathcal{B}$ is a $*$-algebra homomorphism

$$j : \mathcal{B} \to \mathcal{A}.$$

A classical random variable $X : \Omega \to E$ over a probability space $(\Omega, \mathcal{F}, P)$, taking values in a measurable space $(E, \mathcal{E})$, gives rise to a quantum random variable j_X if we take for $\mathcal{B}$ an algebra of complex-valued functions contained in $L^1(E, \mathcal{E}, P_X)$ with P_X the distribution of X and define $j_X(f)$ to be the function $f \circ X$ on Ω. We can take any $\mathcal{A}$ that contains $j_X(\mathcal{B})$. In particular, $\mathcal{B} = L^\infty(E, \mathcal{E})$ and $\mathcal{A} = L^\infty(\Omega, \mathcal{F}, P)$ give a quantum random variable.

The distribution of a quantum random variable j is the state $\varphi_j = \Phi \circ j$ on $\mathcal{B}$. Given a quantum random variable over $(\mathcal{A}, \Phi)$ we always can form the quantum random variable $\pi \circ j$ over the quantum probability space $(\mathbf{H}(\mathcal{D}), \Phi_\Omega)$ where Φ_Ω denotes the vector state on $\mathbf{H}(\mathcal{D})$ given by Ω. Notice that $\pi \circ j$ has the same distribution as j. Let I be a directed set and let $j_i, i \in I$, and j be quantum random variables over $(\mathcal{A}, \Phi)$ on $\mathcal{B}$ with distributions φ_i and φ. The net $(j_i)_{i \in I}$ is said to converge to j in distribution if the net $(\varphi_i(b))_{i \in I}$ of complex numbers converges to $\varphi(b)$ for all $b \in \mathcal{B}$.

A *quantum stochastic process* over a quantum probability space $(\mathcal{A}, \Phi)$ on a $*$-algebra $\mathcal{B}$, indexed by the set T, is a family $(j_t)_{t \in T}$ of quantum random variables

$$j_t : \mathcal{B} \to \mathcal{A};$$

cf. [4]. The states $\varphi_t = \Phi \circ j_t$ are called the *1-dimensional distributions* of (j_t). Two quantum stochastic processes $(j_t^{(i)})_{t \in T}$, $i = 1, 2$, over the quantum probability spaces $(\mathcal{A}^{(i)}, \Phi^{(i)})$ on the same $*$-algebra $\mathcal{B}$ and indexed by the same set T are said to be *equivalent* if

$$\Phi^{(1)}(j_{t_1}^{(1)}(b_1) \ldots j_{t_n}^{(1)}(b_n)) = \Phi^{(2)}(j_{t_1}^{(2)}(b_1) \ldots j_{t_n}^{(2)}(b_n))$$

for all choices of $n \in \mathbb{N}$, $t_1, \dots, t_n \in T$ and $b_1, \dots, b_n \in \mathcal{B}$.

We always can form an equivalent canonical form of a quantum stochastic process as follows. Take $\tilde{\mathcal{A}} = \mathbf{H}(\mathcal{K})$ where $\mathcal{K}$ is the linear span in $\mathcal{D}$ of the set

$$\{\pi \circ j_{t_1}(b_1) \dots j_{t_n}(b_n)\Omega : n \in \mathbb{N}, t_1, \dots, t_n \in T, b_1, \dots, b_n \in \mathcal{B}\},$$

and denote by $\tilde{\Phi}$ the state on $\tilde{\mathcal{A}}$ given by Ω. Then the mappings

$$b \mapsto (\pi \circ j_t(b)) \lceil \mathcal{K}$$

form a quantum stochastic process over $(\tilde{\mathcal{A}}, \tilde{\Phi})$ which is equivalent to the original process $(j_t)_{t \in T}$ on $(\mathcal{A}, \Phi)$. Moreover, two quantum stochastic processes are equivalent if and only if there is a unitary operator

$$\mathrm{U} : \mathcal{H}^{(1)} \to \mathcal{H}^{(2)},$$

$\mathcal{H}^{(i)}$ the completion of $\mathcal{K}^{(i)}$, such that

$$\mathrm{U}\mathcal{K}^{(1)} \subset \mathcal{K}^{(2)}$$

and

$$\mathrm{U}(\pi^{(1)} \circ j_t^{(1)}(b))\xi = (\pi^{(2)} \circ j_t^{(2)}(b))\mathrm{U}\xi$$

for all $t \in T, b \in \mathcal{B}$ and all $\xi \in \mathcal{K}^{(1)}$; see [4].

1.3. Independence. We introduce a notion of independence for n-tuples $(j_1, \dots, j_n)$, $n \in \mathbb{N}$, of quantum random variables over the same quantum probability space $(\mathcal{A}, \Phi)$ and on the same $*$-algebra $\mathcal{B}$. The first condition will be the *factorization property*

$$\Phi(j_{\sigma(1)}(b_1) \dots j_{\sigma(n)}(b_n)) = \varphi_{\sigma(1)}(b_1) \dots \varphi_{\sigma(n)}(b_n) \tag{1.3.1}$$

which is required to hold for all $b_1, \dots, b_n \in \mathcal{B}$ and all elements σ of the permutation group $\mathcal{S}_n$. This condition is borrowed from classical probability theory, and its best justification is the fact that it holds for the various notions of non-commutative independence that appear in the literature; see [19, 23, 38, 47, 68, 82, 88]. In addition, we postulate 'some' *commutation rules* between $j_k(b)$ and $j_l(c)$ for $k \neq l$. For example, this just can be

$$j_k(b) j_l(c) = j_l(c) j_k(b) \tag{1.3.2}$$

for all $k, l = 1, \dots, n$ with $k \neq l$ and all $b, c \in \mathcal{B}$. In this case we speak of *Bose* or *symmetric* independence; see [23]. Another possibility is to generalize (1.3.2) to

$$j_k(b) j_l(c) = \pm j_l(c) j_k(b)$$

where b and c are special elements in $\mathcal{B}$ which can be of degree 0 or 1, and where the sign equals $(-1)^{\mathrm{d}_b \mathrm{d}_c}$. This can be called *Fermi* or *anti-symmetric* independence; see [38].

We generalize these notions of independence. Assume that there are distinguished elements in the algebra $\mathcal{B}$ which obey simpler commutation rules than the other elements. Mathematically, we express this by assuming that $\mathcal{B}$ is a graded $*$-algebra; the distinguished elements are the homogeneous elements. Now suppose that the commutation rules betweeen $j_k(b)$ and $j_l(c)$ for $k < l$ and for homogeneous b in $\mathcal{B}$ are of the form

$$j_k(b)j_l(c) = j_l(\alpha_{\mathrm{d}(b)}c)j_k(b) \tag{1.3.3}$$

where $\alpha_{\mathrm{d}(b)}$ is some mapping on $\mathcal{B}$ that only depends on the degree d_b of b. One can say that c changes to $\alpha_\lambda c$ when a homogeneous element of degree λ passes by. Bose independence is the 'trivial' case $\alpha_\lambda = \mathrm{id}$ for all λ whereas Fermi independence is the case when $\alpha_0 = \mathrm{id}$ and α_1 is the parity operator on the $\mathbb{Z}_2$-graded $*$-algebra $\mathcal{B}$. Equation (1.3.3) is our starting point. Other conditions will follow in a natural way. For instance, the multiplicativity of j_k gives

$$\begin{aligned} j_k(b_1b_2)j_l(c) &= j_k(b_1)j_l(\alpha_{\mathrm{d}(b_2)}c)j_k(b_2) \\ &= j_l(\alpha_{\mathrm{d}(b_1)}\alpha_{\mathrm{d}(b_2)}c)j_k(b_1b_2) \end{aligned}$$

which, on the other hand, equals

$$j_l(\alpha_{\mathrm{d}(b_1)\mathrm{d}(b_2)}c)j_k(b_1b_2).$$

Thus it is natural to postulate

$$\alpha_{\lambda_1}\alpha_{\lambda_2} = \alpha_{\lambda_1\lambda_2} \text{ for all } \lambda_1, \lambda_2 \in \mathbb{L},$$

and, if we extend α linearly to $\mathbb{C}\mathbb{L}$, we have that $\mathcal{B}$ is a module vector space. Similarly, the fact that j_k is a $*$-map yields that $\mathcal{B}$ is a module $*$-vector space. Notice that for $k > l$

$$j_k(b)j_l(c) = j_l(c)j_k(b.\mathrm{d}_c),$$

so that there are commutation rules between all pairs (k, l) with $k \neq l$ (which one would like to have anyway). Next the $*$-algebra homomorphism property of j_l yields that α_λ should be in $\mathbf{Aut}\,\mathcal{B}$ which means that $\mathcal{B}$ actually is a module $*$-algebra. Notice that $\alpha : \mathbb{L} \to \mathbf{Aut}\,\mathcal{B}$ is a $*$-group homomorphism if we equip $\mathbf{Aut}\,\mathcal{B}$ with the $*$-group structure $\alpha \mapsto \alpha^*$ where

$$\alpha^*(a) = \alpha^{-1}(a^*)^*.$$

Finally, we have by associativity

$$\begin{aligned} &j_{k_1}(b_1)j_{k_2}(b_2)j_{k_3}(b_3) \\ &= j_{k_3}((\mathrm{d}(\mathrm{d}_{b_1}.b_2)\mathrm{d}_{b_1}).b_3)j_{k_2}(\mathrm{d}_{b_1}.b_2)j_{k_1}(b_1) \\ &= j_{k_3}((\mathrm{d}_{b_1}\mathrm{d}_{b_2}).b_3)j_{k_2}(\mathrm{d}_{b_1}.b_2)j_{k_1}(b_1) \end{aligned}$$

and this gives

$$\mathrm{d}(\lambda.b)\lambda = \lambda\mathrm{d}_b$$

and $\mathcal{B}$ is a double-module $*$-algebra. Our considerations lead to the following definition. let $\mathcal{B}$ be a double-module $*$-algebra. An n-tuple $(j_1, \ldots, j_n)$ of quantum random variables on $\mathcal{B}$ over a quantum probability space $(\mathcal{A}, \Phi)$ is called *left independent* or (γ, α)-independent if the factorization property (1.3.1) holds and if

$$j_k(b)j_l(c) = j_l(\mathrm{d}_b.c)j_k(b) \tag{1.3.4}$$

for all $b, c \in \mathcal{B}$ with b homogeneous and for all $k < l$. This also can be written

$$M_{\mathcal{A}} \circ (j_k \otimes j_l) = M_{\mathcal{A}} \circ (j_l \otimes j_k) \circ (\alpha \otimes \mathrm{id}) \circ (\mathrm{id} \otimes \tau) \circ (\gamma \otimes \mathrm{id}).$$

If (1.3.4) is replaced by

$$j_k(b) j_l(c) = j_l(c) j_k(b.\mathrm{d}_c)$$

we speak of *right independence* or (α, γ)-independence. If nothing else is said we mean left independence.

Notice that, in general, the order of $j_1, \ldots, j_n$ is important whereas for Bose and Fermi independence it is not. This is why, in the Bose/Fermi case, we have a notion of independence for sets of random variables rather than tuples.

More generally, a family $(j_i)_{i \in I}$ of quantum random variables indexed by a set I with a partial ordering $\preceq$ is called independent if for all choices of $n \in \mathbb{N}$ and $i_1 \prec \cdots \prec i_n$ in I the n-tuple $(j_{i_1}, \ldots, j_{i_n})$ is independent.

1.4. Commutation factors. We introduce an important class of double-module $*$-algebras, namely $*$-algebras with two compatible graduations and with the module structure given by 'commutation factor'.

For two abelian groups $\mathbb{L}$ and $\mathbb{M}$ (written multiplicatively) a $\mathbb{L}$-$\mathbb{M}$-commutation factor is a mapping

$$q : \mathbb{L} \times \mathbb{M} \to \mathbb{C} \setminus \{0\} = \mathbb{C}_0$$

such that

$$q(\lambda, \gamma_1 \gamma_2) = q(\lambda, \gamma_1) q(\lambda, \gamma_2) \tag{1.4.1}$$

$$q(\lambda_1 \lambda_2, \gamma) = q(\lambda_1, \gamma) q(\lambda_2, \gamma) \tag{1.4.2}$$

for all $\lambda, \lambda_1, \lambda_2 \in \mathbb{L}$ and $\gamma, \gamma_1, \gamma_2 \in \mathbb{M}$; see [18]. If $\mathbb{L}$ and $\mathbb{M}$ are $*$-groups we add the condition

$$q(\lambda^*, \gamma^*) = (\overline{q(\lambda, \gamma)})^{-1}. \tag{1.4.3}$$

Given an $\mathbb{L}$-$\mathbb{M}$-commutation factor q we form the sets

$$\mathbb{L}_q^{(0)} = \{\lambda \in \mathbb{L} : q(\lambda, \mu) = 1 \text{ for all } \mu \in \mathbb{M}\}$$

and

$$\mathbb{M}_q^{(0)} = \{\mu \in \mathbb{M} : q(\lambda, \mu) = 1 \text{ for all } \lambda \in \mathbb{L}\}.$$

Then $\mathbb{L}_q^{(0)}$ and $\mathbb{M}_q^{(0)}$ are involutive subgroups of $\mathbb{L}$ and $\mathbb{M}$ respectively and we form the factor groups $\mathbb{L}_q = \mathbb{L}/\mathbb{L}_q^{(0)}$ and $\mathbb{M}_q = \mathbb{M}/\mathbb{M}_q^{(0)}$. Since q gives rise to a $\mathbb{L}_q$-$\mathbb{M}_q$-commutation factor we always can pass to a commutation factor satisfying

$$q(\lambda, \mu) = 1 \text{ for all } \mu \in \mathbb{M} \Rightarrow \lambda = \mathrm{e}$$

$$q(\lambda, \mu) = 1 \text{ for all } \lambda \in \mathbb{L} \Rightarrow \mu = \mathrm{e}$$

The following choice of $\mathbb{L}$ and $\mathbb{M}$ will be the most important for us; it is the case when the commutation rules are given by a complex number not equal to 0. We take $\mathbb{L} = (\mathbb{Z} \times \mathbb{Z})^{\text{twist}}$ and put $\mathbb{M} = \mathbb{Z}^{\text{inv}}$. Notice that $\mathbb{L}$ is nothing but the free abelian involutive semi-group generated by an indeterminate y if we identify $(1,0)$ with y and $(0,1)$ with y^*.

1.4.1. PROPOSITION. *Let $\mathbb{L}$ and $\mathbb{M}$ as above. The equations*

$$q = q((1,0),1); \quad q((m_1,m_2),n) = q^{m_1 n}\,\overline{q}^{m_2 n}$$

establish a 1-1-correspondence between $\mathbb{L}$-$\mathbb{M}$-commutation factors q and non-zero complex numbers (which we again denote by q).

PROOF: We have

$$q((m_1,m_2),n) = q((1,0),1)^{m_1 n} q((0,1),1)^{m_2 n}$$

by conditions (1.4.3) and (1.4.4). Using also (1.4.5), we obtain

$$q((0,1),1)^{-1} = q((0,1),-1) = q((1,0)^*,1^*) = \overline{q((1,0),1)}^{-1}.\diamond$$

Some remarks on special values of q. Denote by $\nu(q)$ the minimum of all numbers l in $\mathbb{N}_0 = \mathbb{N}\setminus\{0\}$ such that $(q/|q|)^l = 1$ if the set of these numbers is not empty. If it is empty we put $\nu(q) = \infty$.
Case 1: $|q| \neq 1, \nu(q) < \infty$, ν odd. Then

$$\mathbb{L}_q = \mathbb{L}/\{(k\nu,-k\nu) : k \in \mathbb{Z}\}$$
$$\mathbb{M}_q = \mathbb{M}.$$

Case 2: $|q| \neq 1$, $\nu(q) < \infty$, ν even. Then

$$\mathbb{L}_q = \mathbb{L}/\{(k\frac{\nu}{2},-k\frac{\nu}{2}) : k \in \mathbb{Z}\}$$
$$\mathbb{M}_q = \mathbb{M}.$$

Case 3: $|q| = 1$. Then

$$\mathbb{L}_q = \mathbb{L}/\{(m,m+k\nu) : m \in \mathbb{Z}\} \cong (\mathbb{Z}_\nu)^{\mathrm{inv}}$$
$$\mathbb{M}_q = \mathbb{M}/\{k\nu : k \in \mathbb{Z}\} = (\mathbb{Z}_\nu)^{\mathrm{inv}}.$$

where $\mathbb{Z}_\infty$ is to be read as $\mathbb{Z}$.
Case 4: None of the cases 1-3: Then $\mathbb{L}_q = \mathbb{L}$ and $\mathbb{M}_q = \mathbb{M}$ as before.

In particular, if q is real, we are in one of the cases 1-3. If $|q| \neq 1$ we have

$$\mathbb{L}_q = \mathbb{L}/\{(k,-k) : k \in \mathbb{Z}\} \cong \mathbb{Z}^{\mathrm{id}}$$
$$\mathbb{M}_q = \mathbb{Z}^{\mathrm{inv}}.$$

If $q = +1$ we have that $\mathbb{L}_q$ and $\mathbb{M}_q$ are trivial. If $q = -1$ we have $\mathbb{L}_q = \mathbb{M}_q = \mathbb{Z}_2$ so that we are in the $\mathbb{Z}_2$-graded case.

A remark on commutation factors for the case when $\mathbb{L}$ and $\mathbb{M}$ are of the form $\mathbb{Z}_\nu$: The only possible involutions on $\mathbb{Z}_\nu$ are the identity and the taking of inverses, and one has to distinguish the two cases that the involutions in $\mathbb{L}$ and $\mathbb{M}$ are of the same or of different type. If the involutions are of the same type the commutation factors are given by complex numbers of modulus 1, and if the involutions are of different type they are given by real numbers not equal to 0. Roughly speaking, the need for two different involutions arises only if one wants to use commutation factors which are not of modulus 1.

A vector space V with an $\mathbb{L}$- and an $\mathbb{M}$-grading is called an $\mathbb{L}$-$\mathbb{M}$-graded vector space if the gradings $(V^{(\lambda)})_{\lambda\in\mathbb{L}}$ and $(V_{(\mu)})_{\mu\in\mathbb{M}}$ are compatible, that is if the $V^{(\lambda)}$ are $\mathbb{M}$-graded and the $V_{(\mu)}$ are $\mathbb{L}$-graded subspaces of V. For an $\mathbb{L}$-$\mathbb{M}$-commutation factor q we define a module structure on V by setting

$$\alpha_\lambda(b) = q(\lambda,\mu)b \tag{1.4.4}$$

for $b \in V_{(\mu)}$. V is a double-module vector space. Similarly, there are $\mathbb{L}$-$\mathbb{M}$-graded $*$-vector spaces, algebras, coalgebras and $*$-algebras which together with a commutation factor yield examples of double-module $*$-vector spaces, algebras, coalgebras and $*$-algebras. For a $\mathbb{L}$-$\mathbb{M}$-commutation factor q an n-tuple $(j_1,\ldots,j_n)$ of quantum random variables over the quantum probability space $(\mathcal{A},\Phi)$ on the $\mathbb{L}$-$\mathbb{M}$-graded $*$-algebra $\mathcal{B}$ is left independent if (1.3.1) holds and if

$$j_k(b)j_l(c) = q(\mathrm{d}(b),\mathrm{g}(c))\; j_l(b)j_k(c)$$

for all k,l, $k<l$, and all homogeneous $b,c\in\mathcal{B}$. Here $\mathrm{d}(b)$ denotes the degree of b with respect to the $\mathbb{L}$-graduation and $\mathrm{g}(c)$ denotes the degree of c with respect to the $\mathbb{M}$-graduation. We speak of left and right q-independence.

1.5. Invariance of states. The factorization property together with the commutation rules in the definition of independence give a restriction of the class of states we are interested in. This does not appear for Bose independence, but already for Fermi independence it means that we are actually dealing with *even* states. We have the following proposition.

1.5.1. PROPOSITION.

(i) *Let (j_1,j_2) be two independent, identically distributed random variables on the double-module $*$-algebra $\mathcal{B}$ and let φ denote the distribution of j_1. Then*

$$\varphi\lceil\mathcal{B}^{(\lambda)} \neq 0 \Rightarrow \varphi(\lambda.b) = \varphi(b) \text{ for all } b\in\mathcal{B}.$$

(ii) *Let q be a commutation factor satisfying*

$$q(\lambda,\mu) = 1 \Rightarrow \lambda = \mathrm{e} \text{ or } \mu = \mathrm{e}. \tag{1.5.1}$$

Moreover, let (j_1,j_2) be q-independent identically distributed random variables on the $\mathbb{L}$-$\mathbb{M}$-graded $*$-algebra $\mathcal{B}$. Then φ is $\mathbb{L}$-even or $\mathbb{M}$-even.

PROOF: (i): For $b\in\mathcal{B}^{(\lambda)}$, $c\in\mathcal{B}$, we have

$$\begin{aligned}\varphi(b)\varphi(c) &= \Phi(j_1(b)j_2(c))\\ &= \Phi(j_2(\lambda.c)j_1(b))\\ &= \varphi(b)\varphi(\lambda.c)\end{aligned}$$

which proves (i).- (ii): Supppose φ is not $\mathbb{L}$-even. Then there exists a $b \in \mathcal{B}$, $\mathrm{d}_b = \lambda \neq \mathrm{e}$, and $\varphi(b) \neq 0$. By (i), for $c \in \mathcal{B}_{(\mu)}$,

$$\varphi(\lambda.c) = q(\lambda, \varepsilon_c)\varphi(c) = \varphi(c).$$

By condition (1.5.1) it follows that $\varepsilon_c \neq \mathrm{e}$ implies $\varphi(c) = 0$.◇

We discuss condition (1.5.1) on q in the case when $q \in \mathbb{C}_0$ and $\mathbb{L} = \mathbb{L}_q$, $\mathbb{M} = \mathbb{M}_q$, like in the discussion following Prop.1.4.1. If q is not of modulus 1 we have that $|q|^{(m_1+m_2)n} = 1$ implies $m_1 = -m_2$ or $n = 0$. If $n \neq 0$ we also have $(q/|q|)^{-2m_1 n} = 1$. This shows that in the case $|q| \neq 1$ condition (1.5.1) is satisfied if and only if either $q/|q| = 1$ or $q/|q|$ is not a root of unity. Assume $|q| = 1$. If q is not a root of unity $\mathbb{L} = \mathbb{M} = \mathbb{Z}^{\mathrm{inv}}$ and (1.5.1) is satisfied. If $q^\nu = 1$, $q^l \neq 1$ for $l < \nu$, we have $\mathbb{L} = \mathbb{M} = (\mathbb{Z}_\nu)^{\mathrm{inv}}$ and it follows from $q^{mn} = 1$ that $mn = 0 \mod \nu$. Thus q has property (1.5.1) if and only if ν is prime. In particular, all real q satisfy condition (1.5.1).

In the sequel, we always assume that our states are *α-invariant*, i.e.

$$\varphi \circ \alpha_\lambda = \varphi \text{ for all } \lambda \in \mathbb{L}.$$

In the commutation factor case this means that φ is $\mathbb{M}$-even.

1.6. Additive and multiplicative white noise. Suppose that we are given a quantum stochastic process (j_{st}) of 'increments' j_{st}. The index set of the process is the set T_{L} of pairs (s,t) of non-negative real numbers with $s \leq t \leq \mathrm{L}$ where L is some fixed real number > 0. We also allow L to be infinity in which case we set $T_\infty = \bigcup_{\mathrm{L}>0} T_{\mathrm{L}}$. We visualize (s,t) as time intervals. If (j_{st}) is a white noise the increments should be independent for disjoint time intervals. T_{L} is a partially ordered set if we define a partial ordering by

$$(s,t) \preceq (s',t') \iff (s,t) = (s',t') \text{ or } t \leq s'.$$

Consider the following axioms for white noise

(WN1) (independence of increments) the family (j_{st}) is independent
(WN2) (stationarity of increments) the distribution of j_{st} only depends on the difference $t - s$
(WN3) (weak continuity) j_{st} converges to j_{ss} in distribution for $t \downarrow s$.

There is one more axiom, namely the one that tells us how 'big' increments are built up from 'small' sub-increments, or to put it in another way, how to combine increments j_{rs} and j_{st} to form the increment j_{rt}. This property, here only vaguely described, will become our fourth axiom, the 'increment property'. Classically, the increment property is governed by the structure of a semi-group G. A classical white noise on a topological semi-group, in our sense, is a stochastic process $(X_{st})_{(s,t)\in T_{\mathrm{L}}}$ taking values in G such that

- $X_{t_1 t_2}, \ldots, X_{t_n t_{n+1}}$ are independent for all choices of $n \in \mathbb{N}$ and $t_1 < \cdots < t_{n+1}$
- X_{st} and $X_{s+r,t+r}$ have the same distribution
- X_{st} converges weakly to X_{ss} for $t \downarrow s$
- $X_{rs}(\omega)X_{st}(\omega) = X_{rs}(\omega)$ almost surely for all $0 \leq r \leq s \leq t$.

If G is a group and $(X_t)_{t\in[0,L]}$ is a G-valued stochastic process with independent, stationary increments $X_{st} = X_s^{-1}X_t$, $s \leq t$, then the 'increment process' $(X_{st})_{(s,t)\in T_L}$ forms a (classical) white noise. In the next section we will see that $*$-bialgebras and $*$-Hopf algebras are the suitable objects to replace semi-groups and groups in the non-commutative case. For the time being, we treat two special cases which are of fundamental interest, namely quantum random vectors and quantum random matrices. The first example is probably the simplest case one can think of. We just assume that the increments are built up 'additively' from smaller sub-increments. In the second example, we start from matrices with non-commuting entries.

Let V be a $*$-vector space. A *quantum random vector* over a probability space $(\mathcal{A}, \Phi)$ is a $*$-mapping $F : V \to \mathcal{A}$. We always can find a vector space basis $\{v_i : i \in I\}$ of V such that the basis elements v_i are hermitian. Then a quantum random vector is a family $(F^{(i)})_{i\in I}$ of hermitian elements $F^{(i)}$ of a quantum probability space. In particular, if V is finite-dimensional, quantum random vectors are tuples $(F^{(1)}, \ldots, F^{(n)})$ of hermitian elements, or, if we apply the GNS-construction, they are tuples of hermitian operators on a pre-Hilbert space and expectations are given by a unit vector of this pre-Hilbert space. In a way, this generalizes the notion of classical $\mathbb{R}^n$-valued random variables, i.e. of classical random vectors. Suppose that $X = (X^{(1)}, \ldots, X^{(n)})$ is a classical random vector over a probability space $(\Omega, \mathcal{F}, P)$, and suppose, furthermore, that all the moments of this random vector are finite. Then $X^{(i)}$, $i = 1, \ldots, n$, act on the linear subspace $\mathcal{D}$ of the Hilbert space $L^2(\Omega, \mathcal{F}, P)$ formed by polynomials in $X^{(i)}$ as linear operators by pointwise multiplication. These operators form a quantum random vector over $(\mathrm{H}(\mathcal{D}), \Phi)$ where Φ is the state given by the function in $\mathcal{D}$ which is constant equal to 1. The underlying involutive vector space is n-dimensional and for a hermitian basis $\{v_i : i = 1, \ldots n\}$ of V we map v_i to the linear operator given by $X^{(i)}$. It is well known (classical moment problem) that, in passing to moments, in general, we loose some information. On the other hand, in many cases, the moments are all one is interested in. Moreover, if the classical random vector is bounded, the corresponding multiplication operators on $L^2(\Omega, \mathcal{F}, P)$ are bounded and we can take the whole Hilbert space as our pre-Hilbert space $\mathcal{D}$. In this case, the classical random vector is completely described by its corresponding quantum random vector.

For a vector space V the tensor algebra $\mathcal{T}(V)$ over V is the vector space $\mathcal{T}(V) = \bigoplus_{n\in\mathbb{N}} V^{\otimes n}$ where $V^{\otimes n}$ denotes the n-fold tensor product of V with itself, $V^{\otimes 0} = \mathbb{C}$, with multiplication given by

$$(v_1 \otimes \ldots \otimes v_n)w = v_1 \otimes \ldots \otimes v \otimes w$$

for $n \in \mathbb{N}$, $v_1, \ldots, v_n, w \in V$. The elements of $\bigcup_{n\in\mathbb{N}} V^{\otimes n}$ are called homogeneous polynomials, and the degree $\mathrm{d}(M)$ of a homogeneous polynomial $M \neq 0$ is n if $M \in V^{\otimes n}$. If $\{v_i : i \in I\}$ is a vector space basis of V the tensor algebra can be viewed as the free algebra generated by the indeterminates $v_i, i \in I$, or as the algebra of all polynomials in the non-commuting indeterminates $v_i, i \in I$. The tensor algebra can also be characterized by the following universal property. There is an embedding of V into $\mathcal{T}(V)$ and for a given linear mapping R from V into an algebra $\mathcal{A}$ there is a unique extension $\mathcal{T}(R)$ of R to the whole of $\mathcal{T}(V)$ such that $\mathcal{T}(R)$ is an algebra homomorphism from $\mathcal{T}(V)$ to $\mathcal{A}$. In a similar way, an involution on V gives rise to a unique extension as an involution on the tensor algebra. Thus for a $*$-vector space V we can form the tensor $*$-algebra $\mathcal{T}(V)$ over V. Moreover, if V is a double-module $*$-vector space, this structure can be extended to the tensor $*$-algebra $\mathcal{T}(V)$ in a unique way such that $\mathcal{T}(V)$ becomes a double-module $*$-algebra.

Now let V be a fixed double-module $*$-vector space. For a quantum random vector F we put $j_F = \mathcal{T}(F)$. Then j_F is a quantum random variable on the double-module $*$-algebra $\mathcal{T}(V)$. Clearly, we have a natural 1-1-correspondence between quantum random vectors and quantum random variables on $\mathcal{T}(V)$. The distribution of the quantum random vector which is identically 0 on V is the state which assigns the value 1 to the unit element $1 \in \mathcal{T}(V)$ and vanishes on all monomials in $\mathcal{T}(V)$ of length > 0. We denote this state by ϵ.

Consider a family $(F_t)_{t\in[0,\mathrm{L}]}$ of quantum random vectors, i.e. a quantum stochastic process on $\mathcal{T}(V)$. For $0 \leq s \leq t \leq \mathrm{L}$ we denote by F_{st} the additive increment $F_t - F_s$. We call (F_t) an *additive white noise* if the family $(j_{st})_{(s,t)\in T_{\mathrm{L}}}$ satisfies the axioms (WN1)-(WN3). Notice that (WN3) can be reformulated to

$$\lim_{t\downarrow 0} \varphi_t(b) = \epsilon(b) \qquad \text{for all } b \in \mathcal{T}(V).$$

Let $(X_t)_{t\in[0,\mathrm{L}]}$ be a classical $\mathbb{R}^d$-valued stochastic process with independent, stationary increments and assume that all moments of the process are finite. Examples are the Wiener and the Poisson process. Then the algebra of functions generated by $\{X_{t,i} : t \in [0,\mathrm{L}], i = 1,\ldots,d\}$ forms a linear subspace $\mathcal{D}$ of $L^2(F,\mathcal{F},P)$. We associate with $X_{t,i}$ the linear operator on $\mathcal{D}$ given by pointwise multiplication with $X_{t,i}$ to obtain an additive white noise.

The following special case of an additive white noise is fundamental. Let V be 2-dimensional, spanned by the vectors x and x^* in V. Let $q \in \mathbb{C}_0$ and $\mathbb{L} = (\mathbb{Z} \times \mathbb{Z})^{\mathrm{twist}}$, $\mathbb{M} = \mathbb{Z}^{\mathrm{inv}}$. A $\mathbb{L}$-$\mathbb{M}$-graduation of $\mathbb{C}\langle x, x^*\rangle$ is defined by setting

$$\mathrm{d}(x) = (1,0), \quad \mathrm{d}(x^*) = (0,1)$$
$$\mathrm{g}(x) = 1, \quad \mathrm{g}(x^*) = -1.$$

This way, V and therefore $\mathcal{T}(V)$, become $\mathbb{L}_q$-$\mathbb{M}_q$-graded with $\mathbb{L}_q$ and $\mathbb{M}_q$ depending on $q \in \mathbb{C}_0$ like in the discussion following Prop. 1.3.1. An additive white noise on V is now a family $(F_t)_{t\in[0,\mathrm{L}]}$ of operators on a pre-Hilbert space $\mathcal{D}$, $F_t \in \mathbf{H}(\mathcal{D})$, and the commutation rules are expressed by

$$F_{st}F_{s't'} = qF_{s't'}F_{st}$$

and

$$F_{st}F^*_{s't'} = q^{-1}F^*_{s't'}F_{st}$$

for $t \leq s'$.

We come two our second type of examples. Let $n \in \mathbb{N}$ be fixed. A *quantum random matrix* over a quantum probability space $(\mathcal{A},\Phi)$ is a $n \times n$-matrix $A = (A_{kl})$ with entries A_{kl} in $\mathcal{A}$. In other words, quantum random matrices are elements of the algebra $\mathrm{M}_n(\mathcal{A})$ with $(\mathcal{A},\Phi)$ a quantum probability space.

Denote by $\mathcal{M}\langle n\rangle$ the free $*$-algebra generated by indeterminates x_{kl}, $k,l = 1,\ldots,n$. Then $\mathcal{M}\langle n\rangle$ is the tensor-$*$-algebra over the involutive vector space spanned by indeterminates x_{kl} and x^*_{kl} with the involution given by $(x_{kl})^* = x^*_{kl}$. For a quantum random matrix A there is a unique $*$-algebra homomorphism

$$\mathcal{T}(A) : \mathcal{M}\langle n\rangle \to \mathcal{A}$$

given by the equation

$$\mathcal{T}(A)(x_{kl}) = A_{kl}.$$

Hence quantum random matrices can be identified with quantum random variables on $\mathcal{M}\langle n\rangle$. A $(\mathbb{Z}\times\mathbb{Z})^{\text{twist}}$-$\mathbb{Z}^{\text{inv}}$-graduation of $\mathcal{M}\langle n\rangle$ is defined by

$$\mathrm{d}(x_{kl}) = (k-l,0),\quad \mathrm{d}(x^*_{kl}) = (0,k-l)$$
$$\mathrm{g}(x_{kl}) = k-l,\quad \mathrm{g}(x^*_{kl}) = l-k.$$

This way we can turn $\mathcal{M}\langle n\rangle$ into a $\mathbb{L}_q$-$\mathbb{M}_q$-graded $*$-algebra with $q\in\mathbb{C}_0$. A family $(A_{st})_{(s,t)\in T_{\mathrm{L}}}$ of quantum random variables over a quantum probability space $(\mathcal{A},\Phi)$ is called a *multiplicative white noise* if the family $(j_{st})_{(s,t)\in T_{\mathrm{L}}}$ satisfies the axioms (WN1)-(WN3) and if, in addition, (A_{st}) is an evolution, that is if

· (evolution property)

$$A_{rs}A_{st} = A_{rt} \qquad \text{for all } r\le s\le t,$$
$$A_{tt} = \mathbf{1}.$$

The commutation rules arise from

$$(A_{st})_{kl}(A_{s't'})_{k'l'} = q^{(k-l)(k'-l')}(A_{s't'})_{k'l'}(A_{st})_{kl}$$

and

$$(A_{st})_{kl}((A_{s't'})^*)_{k'l'} = q^{(k-l)(k'-l')}((A_{s't'})^*)_{k'l'}(A_{st})_{kl}.$$

Suppose that the A_{st} are in the group $\mathrm{GL}_n(\mathcal{A})$ of invertible elements of $\mathrm{M}_n(\mathcal{A})$. Then $A_{st} = A_s^{-1}A_t$, $A_t = A_{0t}$, and a process $(A_t)_{t\in[0,\mathrm{L}]}$, $A_t\in\mathrm{GL}_n(\mathcal{A})$, is called a multiplicative white noise if the multiplicative increments $A_{st} = A_s^{-1}A_t$ form a multiplicative white noise in the above sense.

If the A_t are, in addition, elements of the group $\mathrm{U}_n(\mathcal{A})$ of unitary elements of the $*$-algebra $\mathrm{M}_n(\mathcal{A})$ and if $|q| = 1$ we say that (A_t) is a *unitary q-white noise.* By applying the GNS-construction, a unitary q-white noise $(U_t)_{t\in[0,\mathrm{L}]}$ can always be realized over a quantum probability space of the form $(\mathbf{B}(\mathcal{H}),\Phi)$ with $\mathcal{H}$ a Hilbert space and Φ a state coming from a unit vector in $\mathcal{H}$. The quantum random matrices U_t of a unitary q-white noise then are unitary operators on the Hilbert space $\mathbb{C}^n\otimes\mathcal{H}$. Let $(\Omega,\mathcal{F},P)$ be a classical probability space. As already was pointed out in the introduction of these notes, there is a 1-1-correspondence between classical $\mathrm{U}_n(\mathbb{C})$-valued random variables X over $(\Omega,\mathcal{F},P)$ and elements U of $\mathrm{U}_n(L^\infty(\Omega,\mathcal{F},P))$ given by the equation $U_{kl}(\omega) = X(\omega)_{kl}$. Hence classical unitary random matrices are exactly the quantum random matrices in $\mathrm{U}_n(\mathcal{N})$ over a quantum probability space of the form $(\mathcal{N},\Phi)$ where $\mathcal{N}$ is a commutative von Neumann algebra and Φ is a normal state on $\mathcal{N}$. Thus a classical weakly continuous,

$U_n(\mathbb{C})$-valued stochastic process with independent, stationary multiplicative increments is a symmetric unitary white noise over a quantum probability space $(\mathcal{N}, \Phi)$ of this commutative form.

1.7. Involutive bialgebras. We have seen that additive and multiplicative white noise are analogues of classical stochastic processes with independent, stationary increments taking values in the semi-groups $\mathbb{R}^n$ and M_n respectively. For the analogue of classical white noise we need a non-commutative analogue of the notion of a semi-group. To explain the passage to non-commutativity, we apply a 'Fourier transformation' by, rather than the semi-group G itself, considering the algebra $\mathcal{R}(G)$ of its representative functions which is formed by the coefficients of finite-dimensional representations of G. An involution of the algebra $\mathcal{R}(G)$ is given by complex conjugation of functions. The $*$-algebra $\mathcal{R}(G)$ reflects the right positivity structure, but it does not reflect the semi-group structure of G. Therefore, some additional structure is needed. This is the $*$-bialgebra $\mathcal{R}(G)$ which is the $*$-algebra $\mathcal{R}(G)$ together with the mappings

$$\Delta : \mathcal{R}(G) \to \mathcal{R}(G) \otimes \mathcal{R}(G)$$

and

$$\epsilon : \mathcal{R}(G) \to \mathbb{C}$$

given by $\Delta f(x,y) = f(xy)$ and $\epsilon f = f(\mathrm{e})$ where we identified $\mathcal{R}(G) \otimes \mathcal{R}(G)$ with its image under the bijective linear mapping from $\mathcal{R}(G) \otimes \mathcal{R}(G)$ to $\mathcal{R}(G \times G)$ which maps $f \otimes g$ to the function $(x,y) \mapsto f(x)g(y)$. Both Δ and ϵ are $*$-algebra homomorphisms; see [1] and Section 4.4.

A subcoalgebra $\mathcal{D}$ of a coalgebra $\mathcal{C}$ is a linear subspace of $\mathcal{C}$ such that $\Delta\mathcal{D} \subset \mathcal{D}\otimes\mathcal{D}$. If $\mathcal{D}$ is a subcoalgebra, $(\mathcal{D}, \Delta\lceil\mathcal{D}, \epsilon\lceil\mathcal{D})$ is a coalgebra. The intersection and the sum $\mathcal{D}_1 + \mathcal{D}_2 = \{d_1 + d_2 : d_1 \in \mathcal{D}_1, d_2 \in \mathcal{D}_2\}$ of two subcoalgebras $\mathcal{D}_1$ and $\mathcal{D}_2$ are again subcoalgebras. For a subset X of $\mathcal{C}$ we denote by $\mathcal{C}_X$ the smallest subcoalgebra containing X. For a finite-dimensional linear subspace V of a coalgebra $\mathcal{C}$ the subcoalgebra $\mathcal{C}_V$ is finite-dimensional by the *fundamental theorem on coalgebras*; see [87]. A proof of this theorem can also be found in Section 4.4 of these notes. It follows that a coalgebra is the inductive limit of the system $\mathcal{U}(\mathcal{C})$ of its finite-dimensional subcoalgebras. Frequently we make use of the following construction. Let $\mathcal{C}$ be a coalgebra and let W be a vector space. Let $(R_\mathcal{D})_{\mathcal{D}\in\mathcal{U}(\mathcal{C})}$ be a family of linear mappings $R_\mathcal{D} : \mathcal{D} \to W$ satisfying the consistency condition

$$R_\mathcal{D}\lceil\mathcal{D}' = R_{\mathcal{D}'} \text{ for } \mathcal{D}, \mathcal{D}' \in \mathcal{U}(\mathcal{C}) \text{ with } \mathcal{D}' \subset \mathcal{D}.$$

Then there is a uniquely determined linear map $R : \mathcal{C} \to W$, called the inductive limit of $(R_\mathcal{D})_{\mathcal{D}\in\mathcal{U}(\mathcal{C})}$, such that

$$R\lceil\mathcal{D} = R_\mathcal{D} \text{ for } \mathcal{D} \in \mathcal{U}(\mathcal{C}).$$

In particular, if S is a linear operator on $\mathcal{C}$ leaving invariant all subcoalgebras of $\mathcal{C}$ we define the linear operator $\exp\, S$ on $\mathcal{C}$ to be the inductive limit of the system $(\exp(S\lceil\mathcal{D}))_{\mathcal{D}\in\mathcal{U}(\mathcal{C})}$. Moreover, we can define $\exp(\mathrm{ad}\, S)$ as an operator on $\mathbf{L}(\mathcal{C})$ as follows. For $\mathcal{D}, \mathcal{D}' \in \mathcal{U}(\mathcal{D})$ we have that $\mathrm{ad}\, S$ leaves $\mathbf{L}(\mathcal{D}, \mathcal{D}')$ invariant. Thus we can define $\exp(\mathrm{ad}\, S)$ as an operator on $\mathbf{L}(\mathcal{D}, \mathcal{D}')$. For $T \in \mathbf{L}(\mathcal{C})$ and $\mathcal{D} \in \mathcal{U}(\mathcal{C})$ let $R_\mathcal{D}$ be the linear map $\exp(\text{ad } S)(T)$ from $\mathcal{D}$ to $\mathcal{C}_{T(\mathcal{D})}$. Define $\exp(\text{ad } S)(T)$ to be the inductive limit of $(R_{\mathcal{D}})_{\mathcal{D}\in\mathcal{U}(\mathcal{C})}$. We have the relation

$$(\exp S) \circ T \circ (\exp(-S)) = \exp(\text{ad } S)(T). \tag{1.7.1}$$

Let $(\mathcal{C}, \Delta, \epsilon)$ be a coalgebra and let $(\mathcal{A}, M, \kappa)$ be an algebra. The vector space $\mathbf{L}(\mathcal{C}, \mathcal{A})$ is turned into an algebra if we set

$$R \star S = M \circ (R \otimes S) \circ \Delta$$

for $R, S \in \mathbf{L}(\mathcal{C}, \mathcal{A})$. The unit element of this algebra is given by $b \mapsto \epsilon(b)\mathbf{1}$. We call $R \star S$ the convolution product of R with S. In particular, if we take $\mathcal{A} = \mathbb{C}$ the algebraic dual space $\mathcal{C}^*$ of $\mathcal{C}$ becomes an algebra. For $\psi \in \mathcal{C}^*$ we define $\mathcal{R}(\psi)$ to be the linear operator $(\text{id} \otimes \psi) \circ \Delta$ on $\mathcal{C}$. Then $\mathcal{R} : \mathcal{C}^* \to \mathbf{L}(\mathcal{C})$ is an algebra homomorphism where we consider $\mathbf{L}(\mathcal{C})$ as the algebra of linear operators on $\mathcal{C}$ with the multiplication given by composition. Moreover, $\epsilon \circ \mathcal{R}(\psi) = \psi$. Clearly, $\mathcal{R}(\psi)$ leaves invariant each subcoalgebra of $\mathcal{C}$ and we can define the convolution exponential $\exp_\star \psi \in \mathcal{C}^*$ of ψ by setting

$$(\exp_\star \psi)(b) = (\epsilon \circ \exp\{\mathcal{R}(\psi)\})(b).$$

Let V be a comodule vector space and let W be a module vector space. We define the twist map

$$\tau_{(\gamma,\alpha)} : V \otimes W \to W \otimes V$$

to be the linear mapping given by

$$\tau_{(\gamma,\alpha)}(v \otimes w) = \mathrm{d}_v.w \otimes v,$$

$v \in V$, $w \in W$. In the commutation factor case we write τ_q for $\tau_{(\gamma,\alpha)}$. If V and W are involutive we have that

$$v \otimes w \mapsto \tau^{-1}_{(\gamma,\alpha)}(w^* \otimes v^*) = v^* \otimes w^*.\mathrm{d}^*_v$$

defines an involution on the vector space $V \otimes W$. If V and W both are double-module $*$-vector spaces then $V \otimes W$ is again a double-module $*$-vector space; the module structure is given by

$$\lambda.(v \otimes w) = \lambda.v \otimes \lambda.w.$$

We denote it by $V \otimes_{(\gamma,\alpha)} W$. A coalgebra $(\mathcal{C}, \Delta, \epsilon)$ is called an involutive (γ, α)-coalgebra if it is a double-module coalgebra and if $\mathcal{C}$ is a double-module $*$-vector space such that Δ is a $*$-mapping from $\mathcal{C}$ to $\mathcal{C} \otimes_{(\gamma,\alpha)} \mathcal{C}$.

Notice that, although the notion of a coalgebra is dual to the notion of an algebra, this relationship does not hold between $*$-coalgebras and graded $*$-algebras; cf. also the remark following the next proposition. This is also reflected in the fact that the dual algebra $\mathcal{C}^*$ of a $*$-coalgebra, in general, cannot be turned into a $*$-algebra in the natural way. Indeed we have for $\psi, \varphi \in \mathcal{C}^*$

$$(\psi \star \varphi)^* = \psi^* \star \varphi^*$$

and $\psi \mapsto \psi^*$ is an antilinear homomorphism but not an involution unless $\mathcal{C}^*$ is commutative. We can form tensor products.

1.7.1. PROPOSITION.

(i) *For a comodule coalgebra* $(\mathcal{C}, \Delta, \epsilon)$ *and a module coalgebra* $(\mathcal{D}, \Pi, \pi)$ *the triplet* $(\mathcal{C} \otimes \mathcal{D}, \Delta \otimes_{(\gamma,\alpha)} \Pi, \epsilon \otimes \pi)$ *with*

$$\Delta \otimes_{(\gamma,\alpha)} \Pi = (\mathrm{id} \otimes \tau_{(\gamma,\alpha)} \otimes \mathrm{id}) \circ (\Delta \otimes \Pi)$$

is a coalgebra. It is called the (γ, α)*-tensor product of the coalgebras* $\mathcal{C}$ *and* $\mathcal{D}$ *and is denoted by* $\mathcal{C} \otimes_{(\gamma,\alpha)} \mathcal{D}$.

(ii) *For a comodule algebra* $(\mathcal{A}, M, \kappa)$ *and a module algebra* $(\mathcal{B}, N, \nu)$ *the triplet* $(\mathcal{A} \otimes \mathcal{B}, M \otimes_{(\gamma,\alpha)} N, \kappa \otimes \nu)$ *with*

$$M \otimes_{(\gamma,\alpha)} N = (M \otimes N) \circ (\mathrm{id} \otimes \tau^{-1}_{(\gamma,\alpha)} \otimes \mathrm{id})$$

is an algebra. It is called the (γ, α)*-tensor product of the algebras* $\mathcal{A}$ *and* $\mathcal{B}$ *and is denoted by* $\mathcal{A} \otimes_{(\gamma,\alpha)} \mathcal{B}$. *If* $\mathcal{A}$ *and* $\mathcal{B}$ *both are double-module* $*$*-algebras then* $\mathcal{A} \otimes_{(\gamma,\alpha)} \mathcal{B}$ *is again a double-module* $*$*-algebra.*

PROOF: Repeated use of the properties of module and comodule algebras and coalgebras.◇

Similarly, (α, γ)-tensor products can be defined. In general, there is no analogue of the statement in (ii) on involutions for $*$-coalgebras. The (γ, α)-tensor product of two coalgebras which are also double-module $*$-coalgebras is not necessarily again a double-module $*$-coalgebra. However, we have

$$((\Delta \otimes_{(\gamma,\alpha)} \Pi)(c \otimes d))^* = (\Delta \otimes_{(\alpha,\gamma)} \Pi)((c \otimes d)^*)$$

where the involution on the left hand side it is the one in $(\mathcal{C} \otimes_{(\gamma,\alpha)} \mathcal{D}) \otimes_{(\gamma,\alpha)} (\mathcal{C} \otimes_{(\gamma,\alpha)} \mathcal{D})$ and on the right hand side is the one in $\mathcal{C} \otimes_{(\gamma,\alpha)} \mathcal{D}$.

The space $\mathbf{L}(V)$ with V a double-module vector space is an example of a double-module algebra. The module structure is given by

$$(\lambda.R)(v) = \lambda.R(v.\lambda).$$

For double-module vector spaces V and W we have that $\mathbf{L}(V, W)$ and $V \otimes W$ are again double-module vector spaces in a natural way. If V_1, V_2, W_1 and W_2 are double-module vector spaces an embedding Ψ of $\mathbf{L}(V_1, V_2) \otimes \mathbf{L}(W_1, W_2)$ into the double-module vector space $\mathbf{L}(V_1 \otimes W_1, V_2 \otimes W_2)$ is given by

$$\Psi(R \otimes T)(v \otimes w) = Rv \otimes (T.\mathrm{d}_v)w$$

for $R \in \mathbf{L}(V_1, V_2)$, $T \in \mathbf{L}(W_1, W_2)$, $v \in V_1$ and $w \in W_1$. In this way, $\mathbf{L}(V) \otimes \mathbf{L}(W)$ is considered as a double-moduled subalgebra of $\mathbf{L}(V \otimes W)$ for double-module vector spaces V and W. This double-module algebra structure on $\mathbf{L}(V) \otimes \mathbf{L}(W)$ coincides with $\mathbf{L}(V) \otimes_{(\gamma,\alpha)} \mathbf{L}(W)$.

The following can easily be verified.

1.7.2. PROPOSITION.

(i) *For a double-module $*$-vector space V the twist map $\tau_{(\gamma,\alpha)}$ is a $*$-map from $V \otimes_{(\gamma,\alpha)} V$ to $V \otimes_{(\alpha,\gamma)} V$.*

(ii) *For a double-module coalgebra $\mathcal{C}$ the twist map $\tau_{(\gamma,\alpha)}$ is a coalgebra homomorphisms from $\mathcal{C} \otimes_{(\alpha,\gamma)} \mathcal{C}$ to $\mathcal{C} \otimes_{(\gamma,\alpha)} \mathcal{C}$.*

(iii) *For a double-module algebra $\mathcal{A}$ the twist map $\tau_{(\gamma,\alpha)}$ is an algebra homomorphism from $\mathcal{A} \otimes_{(\gamma,\alpha)} \mathcal{A}$ to $\mathcal{A} \otimes_{(\alpha,\gamma)} \mathcal{A}$.*$\diamondsuit$

A $\mathbb{Z}_2$-graded coalgebra $(\mathcal{C}, \Delta, \epsilon)$ is called *cocommutative* if

$$\tau_{-1} \circ \Delta = \Delta. \tag{1.7.2}$$

It is not obvious what cocommutativity should mean in the general case of a double-module coalgebra.

Let $\mathcal{B}$ be a double-module $*$-algebra. The additional structure on $\mathcal{B}$ that replaces the group structure appearing in the notion of classical white noise is the structure of a coalgebra. Thus let us suppose that $\mathcal{B}$ also is a coalgebra. We assume that $\mathcal{B}$ is a double-module coalgebra. This guarantees that the convolution product of α-invariant mappings (states for example) is again α-invariant and that the convolution product of even mappings is even again. We derive a compatibility condition between the $*$-algebra and the coalgebra structure. Since the convolution product $j_1 \star j_2$ of a pair of (j_1, j_2) of independent random variables should again be a random variable, we must have

$$\begin{aligned}(j_1 \star j_2)(bc) &= \sum j_1((bc)_{(1)}) j_2((bc)_{(2)}) \\ &= \sum j_1(b_{(1)}) j_2(b_{(2)}) j_1(c_{(1)}) j_2(c_{(2)}) \\ &= \sum j_1(b_{(1)}c_{(1)}) j_2((b_{(2)}.\mathrm{d}(c_{(1)}))c_{(2)})\end{aligned}$$

and

$$\begin{aligned}(j_1 \star j_2)(b^*) &= \sum j_1((b^*)_{(1)}) j_2((b^*)_{(2)}) \\ &= \sum j_2(b_{(2)}{}^*) j_1(b_{(1)}{}^*) \\ &= \sum j_1(b_{(1)}{}^*) j_2(b_{(2)}{}^*.\mathrm{d}(b_{(1)})^*)\end{aligned}$$

which gives

$$\Delta(bc) = \sum b_{(1)}c_{(1)} \otimes (b_{(2)}.\mathrm{d}(c_{(1)}))c_{(2)}$$

and

$$\Delta(b^*) = \sum b_{(1)}{}^* \otimes b_{(2)}{}^*.\mathrm{d}(b_{(1)})^*.$$

A (γ, α)-*bialgebra* $\mathcal{B}$ is a double-module coalgebra $(\mathcal{B}, \Delta, \epsilon)$ which is also a double-module algebra $(\mathcal{B}, M, \kappa)$ in such a way that

$$\Delta : \mathcal{B} \to \mathcal{B} \otimes_{(\gamma,\alpha)} \mathcal{B} \text{ and } \epsilon : \mathcal{B} \to \mathbb{C}$$

are algebra homomorphisms. This is easily seen to be equivalent to the condition that

$$M : \mathcal{B} \otimes_{(\alpha,\gamma)} \mathcal{B} \to \mathcal{B} \text{ and } \kappa : \mathbb{C} \to \mathcal{B}$$

are coalgebra homomorphisms. An *involutive* (γ, α)*-bialgebra* $\mathcal{B}$ is a (γ, α)-bialgebra such that $\mathcal{B}$ is also a double-module $*$-vector space which turns $(\mathcal{B}, \Delta, \epsilon)$ into an involutive (γ, α)-coalgebra and $(\mathcal{A}, M, \kappa)$ into a double-module $*$-algebra. In the commutation factor case we speak of (left and right) q-bialgebras and so on. In the trivial graduation (i.e. $q = 1$) case we speak of *symmetric* bialgebras.

The opposite of an involutive (γ, α)-bialgebra $\mathcal{B}$ is the involutive (α, γ)-bialgebra $\mathcal{B}^{\mathrm{op}}$ which equals $\mathcal{B}$ as a double-module $*$-algebra but as a coalgebra is $(\mathcal{B}, \Delta^{\mathrm{op}}, \epsilon)$ where

$$\Delta^{\mathrm{op}} = \mathcal{T}_{(\gamma,\alpha)} \circ \Delta.$$

A (γ, α)-bialgebra $\mathcal{B}$ is called a (γ, α)*-Hopf algebra* if there is an even, α-invariant linear mapping S on $\mathcal{B}$, called the *antipode* of $\mathcal{B}$, which is the inverse of id with respect to the convolution algebra structure of $\mathbf{L}(\mathcal{B})$, i.e.

$$M \circ (S \otimes \mathrm{id}) \circ \Delta = M \circ (\mathrm{id} \otimes S) \circ \Delta = \epsilon \mathbf{1}.$$

An involutive (γ, α)-Hopf algebra is a (γ, α)-Hopf algebra which also is an involutive (γ, α)-bialgebra plus the compatibility condition $S(S(b^*)^*) = b$ for the antipode. We have

1.7.3. PROPOSITION. *Let $\mathcal{B}$ be a (γ, α)-Hopf algebra with antipode S. Then*

$$S(bc) = S(c)S(b.\mathrm{d}_c).$$

PROOF: This follows from the fact that both

$$b \otimes c \mapsto S(bc)$$

and

$$b \otimes c \mapsto S(c)S(b.\mathrm{d}_c)$$

define an inverse of the multiplication in $\mathcal{B}$ with respect to the convolution algebra structure of $L(\mathcal{B} \otimes_{(\gamma,\alpha)} \mathcal{B}, \mathcal{B})$.$\diamondsuit$

1.8. Examples. An element b of a coalgebra $\mathcal{C}$ is called *group-like*, or *exponential*, if

$$\Delta b = b \otimes b.$$

The counit property forces $\epsilon b = 1$. An element b of a coalgebra $\mathcal{C}$ is called (c, d)*-primitive*, $c, d \in \mathcal{C}$, if

$$\Delta b = c \otimes b + b \otimes d.$$

If c and d both are group-like, then for a (c, d)-primitive element b we must have $\epsilon b = 0$ by the counit property. A $(\mathbf{1}, \mathbf{1})$-primitive element of a (γ, α)-bialgebra is called a *Lie element.*

1. SEMI-GROUP ALGEBRAS: Let G be a semi-group. Then $\mathbb{C}G$ becomes a symmetric bialgebra if we define the elements of G to be group-like and extend linearly. This

bialgebra is cocommutative and it is commutative if and only if G is commutative. If G is a $*$-semi-group $\mathbb{C}G$ becomes a $*$-bialgebra, and if G is a group $\mathbb{C}G$ becomes a Hopf algebra with the antipode given by linear extension of $g \mapsto g^{-1}$. If G is a $*$-group we define an involution on $\mathbb{C}G$ by anti-linear extension of $g \mapsto g^*$ to obtain a $*$-Hopf algebra.

2. Tensor algebras: Let V be a double-module $*$-vector space. The double-module tensor $*$-algebra $\mathcal{T}(V)$ over V becomes an involutive (γ, α)-Hopf algebra if we define the elements of V to be Lie elements and extend

$$\Delta : \mathcal{T}(V) \to \mathcal{T}(V) \otimes_{(\gamma,\alpha)} \mathcal{T}(V)$$

and

$$\epsilon : \mathcal{T}(V) \to \mathbb{C}$$

to $*$-algebra homomorphisms. The antipode of this (γ, α)-Hopf algebra is given by extension of $v \mapsto -v$, $v \in V$. In the commutation factor case, we sometimes write $\mathcal{T}(V; q)$ for this structure to indicate the dependence of the comultiplication on q. In the case when $\mathbb{L} = \mathbb{M} = \mathbb{Z}_2$, the $\mathbb{Z}_2$-graded $*$-Hopf algebra $\mathcal{T}(V)$ is also cocommutative.

3. Enveloping Hopf algebras: Let $\mathbf{g}$ be a complex $\mathbb{Z}_2$-graded Lie algebra. The universal enveloping algebra $\mathrm{Env}(\mathbf{g})$ of $\mathbf{g}$ is the tensor algebra $\mathcal{T}(\mathbf{g})$ divided by the ideal in $\mathcal{T}(\mathbf{g})$ generated by the elements

$$v \otimes w - (-1)^{\mathrm{d}(v)\mathrm{d}(w)} w \otimes w - [v, w],$$

$v, w \in \mathbf{g}$. One checks that this ideal is a graded coideal of the $\mathbb{Z}_2$-graded bialgebra $\mathcal{T}(\mathbf{g})$. Thus $\mathrm{Env}(\mathbf{g})$ becomes a $\mathbb{Z}_2$-graded (cocommutative) Hopf algebra. If $\mathbf{g}$ has an involution, that is $\mathbf{g}$ is an involutive vector space such that

$$[v, w]^* = [w^*, v^*] \text{ for all } v, w \in \mathbf{g},$$

then $\mathrm{Env}(\mathbf{g})$ is a $\mathbb{Z}_2$-graded $*$-Hopf algebra. The $\mathbb{Z}_2$-graded Lie algebras obtained from $\mathbb{Z}_2$-graded $*$-algebras are examples of graded Lie $*$-algebras. The complexification $\mathbf{g}_{\mathbb{C}} = \mathbf{g} \oplus i\mathbf{g}$ of a real graded Lie algebra $\mathbf{g}$ with $v^* = -v$ for $v \in \mathbf{g}$ is another example.

4. Coefficient algebras of matrix semi-groups: In the beginning of the foregoing section we introduced the algebra $\mathcal{R}(G)$ of representative functions of a semi-group G as a motivating example of a $*$-bialgebra. Roughly speaking, $\mathcal{R}(G)$ is the dual of the semi-group algebra in the sense that what was multiplication in $\mathcal{R}(G)$ is comultiplication in $\mathbb{C}G$ and vice versa. If G is a topological semi-group we denote by $\mathcal{R}_0(G)$ the sub-$*$-bialgebra of $\mathcal{R}(G)$ formed by the continuous representative functions. If G is a group then $\mathcal{R}(G)$ and $\mathcal{R}_0(G)$ are $*$-Hopf algebras with the antipode $(Sf)(x) = f(x^{-1})$.

For a symmetric $*$-bialgebra $\mathcal{B}$ we denote by $\mathbf{A}(\mathcal{B})$ the set of all $*$-algebra homomorphisms from $\mathcal{B}$ to $\mathbb{C}$. Then $\mathbf{A}(\mathcal{B})$ is a convolution sub-semi-group of $\mathcal{B}^*$, and if $\mathcal{B}$ is a $*$-Hopf algebra, $\mathbf{A}(\mathcal{B})$ is a group with the inverse of $\chi \in \mathbf{A}(\mathcal{B})$ given by $\chi^{-1}(b) = (\chi \circ S)(b)$. In the case when the group G is locally compact abelian or compact we have $\mathbf{A}(\mathcal{R}_0(G)) \cong G$ which is the algebraic part of the Pontrjagin and the Krein-Tanaka duality theory respectively; see e.g. [34].

In particular, if G is the group $\mathrm{U}(n)$ of unitary $n\times n$-matrices one can show that $\mathcal{R}_0(G)$ is isomorphic to the commutative polynomial algebra $\mathcal{M}[n]$ in n^2 indeterminates x_{kl}, x^*_{kl}, $k,l=1,\ldots,n$, divided by the ideal $\mathcal{I}$ in $\mathcal{M}[n]$ generated by the elements

$$\sum_{m=1}^{n} x_{km}x^*_{lm} - \delta_{kl} \text{ and } \sum_{m=1}^{n} x^*_{mk}x_{ml} - \delta_{kl}; \tag{1.8.1}$$

see e.g. [30]. The involution is given by $(x_{kl})^* = x^*_{kl}$ and the comultiplication and counit are obtained by first extending

$$\Delta : \mathcal{M}[n] \to \mathcal{M}[n]\otimes\mathcal{M}[n]; \quad \epsilon : \mathcal{M}[n] \to \mathbb{C}$$

with

$$\Delta x_{kl} = \sum_{m=1}^{n} x_{km}\otimes x_{ml}; \quad \epsilon x_{kl} = \delta_{kl} \tag{1.8.2}$$

to a $*$-algebra homomorphism, turning $\mathcal{M}[n]$ into a symmetric $*$-bialgebra, and then showing that the ideal $\mathcal{I}$ is a $*$-coideal. The antipode of $\mathcal{U}[n] = \mathcal{R}_0(\mathrm{U}(n)) = \mathcal{M}[n]/\mathcal{I}$ is given by $x_{kl}\mapsto x^*_{lk}$. A sub-$*$-Hopf algebra of $\mathcal{U}[2]$ is $\mathcal{SU}[2] = \mathcal{R}_0(\mathrm{SU}(2))$ which is obtained by dividing $\mathcal{U}[2]$ by the ideal generated by the elements $x^*_{11}-x_{22}$ and $x^*_{12}-x_{21}$.

Another commutative symmetric $*$-Hopf algebra is defined as follows. We denote by $\mathcal{H}[2]$ the commutative $*$-algebra generated by the indeterminates x, y, y^{-1}, z and z^{-1} with the relations $yy^{-1} = zz^{-1} = \mathbf{1}$. We turn $\mathcal{H}[2]$ into a symmetric $*$-bialgebra by defining z and y to be group-like and x to be (z,y)-primitive. We also are interested in the quotient $\mathcal{H}^{(0)}[2]$ of $\mathcal{H}[2]$ by the $*$-ideal generated by the element $z-\mathbf{1}$. Then $\mathcal{H}^{(0)}[2]$ is the commutative $*$-algebra generated by x, y and y^{-1} with the relation $yy^{-1} = \mathbf{1}$ and such that y is group-like and x is $(\mathbf{1},y)$-primitive. Finally, $\mathcal{H}^{(00)}[2]$ is defined as the sub-$*$-bialgebra of $\mathcal{H}^{(0)}[2]$ generated by x and y. An antipode of $\mathcal{H}[2]$ is given by $x\mapsto -z^{-1}xy^{-1}$, $x^*\mapsto -(z^*)^{-1}x^*(y^*)^{-1}$, so that $\mathcal{H}[2]$ and $\mathcal{H}^{(0)}[2]$ are $*$-Hopf algebras. We have that $\mathbf{A}(\mathcal{H}[2])$ is the Heisenberg group $\mathrm{H}(2)$ formed by the 2×2-matrices of the form $\begin{pmatrix}\alpha & \beta\\ 0 & \gamma\end{pmatrix}$, $\alpha,\gamma\in\mathbb{C}_0$, $\beta\in\mathbb{C}$. Moreover, $\mathbf{A}(\mathcal{H}^{(0)}[2])$ is the subgroup $\mathrm{H}^{(0)}(2)$ of $\mathrm{H}(2)$ formed by the matrices $\begin{pmatrix}1 & \beta\\ 0 & \gamma\end{pmatrix}$, $\gamma\in\mathbb{C}_0$, $\beta\in\mathbb{C}$, and $\mathbf{A}(\mathcal{H}^{(00)}[2])$ is the semi-group $\mathrm{H}^{(00)}(2)$ of all complex 2×2-matrices of the form $\begin{pmatrix}1 & \beta\\ 0 & \gamma\end{pmatrix}$. We say that $\mathcal{U}[n]$, $\mathcal{SU}[2]$, $\mathcal{H}[2]$, $\mathcal{H}^{(0)}[2]$ and $\mathcal{H}^{(00)}[2]$ are coefficient algebras of $\mathrm{U}(n), \mathrm{SU}(2), \mathrm{H}(2), \mathrm{H}^{(0)}(2)$ and $\mathrm{H}^{(00)}(2)$ respectively.

5. Non-commutative analogues and deformations of coefficient algebras: The non-commutative analogue of $\mathcal{M}[n]$ is the $*$-algebra $\mathcal{M}\langle n\rangle$ generated by *non-commuting* indeterminates x_{kl}, $k,l=1,\ldots,n$, which was already introduced in Section 1.6. We have seen that, for a given commutation factor $q\in\mathbb{C}_0$, the $*$-algebra $\mathcal{M}\langle n\rangle$ can be turned into a $(\mathbb{Z}\times\mathbb{Z})^{\mathrm{twist}}$-$\mathbb{Z}^{\mathrm{inv}}$-graded $*$-algebra. If we define

$$\Delta : \mathcal{M}\langle n\rangle \to \mathcal{M}\langle n\rangle\otimes_q\mathcal{M}\langle n\rangle; \quad \epsilon : \mathcal{M}\langle n\rangle \to \mathbb{C}$$

again by extension of (1.8.2) we obtain an involutive q-bialgebra which we denote by $\mathcal{M}\langle n;q\rangle$. Next we have

1.8.1. PROPOSITION. *Let $|q|=1$. Then the two-sided ideal $\mathcal{I}$ in $\mathcal{M}\langle n;q\rangle$ generated by the elements of the form (1.8.1) is a $\mathbb{L}_q$-$\mathbb{M}_q$-graded $*$-coideal.*

PROOF: In the case $|q|=1$ we have $\mathbb{L}_q=\mathbb{M}_q=(\mathbb{Z}_q)^{\mathrm{inv}}$, so that the elements (1.8.1) are homogeneous. Thus $\mathcal{I}$ is a $\mathbb{L}_q$-$\mathbb{M}_q$-graded $*$-ideal. Clearly, $\mathcal{I}$ is a graded $*$-ideal. Moreover,

$$
\begin{aligned}
\Delta(\sum_{m=1}^{n} x_{km}x_{lm}^{*}) &= \sum_{m,r,s=1}^{n}(x_{kr}\otimes x_{rm})(x_{ls}\otimes x_{sm})^{*} \\
&= \sum_{m,r,s=1}^{n}\bar{q}^{(l-s)(s-m)+(r-m)(s-l)}x_{kr}x_{ls}^{*}\otimes x_{rm}x_{sm}^{*} \\
&= \sum_{r,s=1}^{n}\bar{q}^{(l-s)(s-r)}x_{kr}x_{ls}^{*}\otimes(\sum_{m=1}^{n}x_{rm}x_{sm}^{*}-\delta_{rs}\mathbf{1}) \\
&\quad +(\sum_{r=1}^{d}x_{kr}x_{lr}^{*}-\delta_{kl}\mathbf{1})\otimes\mathbf{1}+\delta_{kl}\otimes\mathbf{1}
\end{aligned}
$$

which shows that

$$
\Delta(\sum_{m=1}^{n}x_{km}x_{lm}^{*}-\delta_{kl}\mathbf{1})\subset\mathcal{I}\otimes M_{\mathbb{C}}\langle n\rangle+M_{\mathbb{C}}\langle n\rangle\otimes\mathcal{I}.
$$

A similar calculation holds for the other type of generators of $\mathcal{I}$.◇

For $|q|=1$ we define the non-commutative analogue $\mathcal{U}\langle n;q\rangle$ of $\mathcal{U}[n]$ to be the quotient of $\mathcal{M}\langle n;q\rangle$ by the ideal $\mathcal{I}$. Then $\mathcal{U}\langle n;q\rangle$ is an involutive q-bialgebra, but $\mathcal{U}\langle n;q\rangle$ is not a Hopf algebra. The $*$-algebra $\mathcal{U}\langle n\rangle$ is isomorphic to the $*$-algebra generated by the mappings $U\mapsto U_{kl}$, $k,l=1,\dots,n$, from the group $\mathrm{U}(\mathbf{B}(\mathbb{C}^n\otimes\mathcal{H}))$ to the $*$-algebra $\mathbf{B}(\mathcal{H})$ where $\mathcal{H}$ is an infinite-dimensional Hilbert space and where the multiplication and the involution are given by the pointwise multiplication and by the pointwise taking of adjoints in $\mathbf{B}(\mathcal{H})$; see [30].

Next $\mathcal{SU}\langle 2;q\rangle$ is defined as the quotient of $\mathcal{U}\langle 2;q\rangle$ by the relations $x_{11}^{*}=x_{22}$ and $x_{12}^{*}=x_{21}$.

The $*$-algebras $\mathcal{H}\langle 2\rangle$, $\mathcal{H}^{(0)}\langle 2\rangle$ and $\mathcal{H}^{(00)}\langle 2\rangle$ are defined in a now obvious way. Given $q\in\mathbb{C}\setminus\{0\}$ we turn $\mathcal{H}\langle 2\rangle$ into a $(\mathbb{Z}\times\mathbb{Z})^{\mathsf{twist}}$-$\mathbb{Z}^{\mathrm{inv}}$-graded $*$-algebra by taking over the graduations of the tensor-$*$-algebra $\mathbb{C}\langle x,x^{*}\rangle\subset\mathcal{H}\langle 2\rangle$ and putting the degrees of z,z^{-1},y and y^{-1} all equal to 0. Then $\mathcal{H}\langle 2\rangle$ is turned into an involutive q-bialgebra by defining z and y to be group-like and x to be (z,y)-primitive and by extending Δ and ϵ to $*$-algebra homomorphisms

$$
\Delta:\mathcal{H}\langle 2\rangle\to\mathcal{H}\langle 2\rangle\otimes_q\mathcal{H}\langle 2\rangle;\quad \epsilon:\mathcal{H}\langle 2\rangle\to\mathbb{C}.
$$

This way, $\mathcal{H}\langle 2\rangle$ and $\mathcal{H}^{(0)}\langle 2\rangle$ become involutive q-Hopf algebras and $\mathcal{H}^{(00)}\langle 2\rangle$ becomes an involutive q-bialgebra which we denote by $\mathcal{H}\langle 2;q\rangle$, $\mathcal{H}^{(0)}\langle 2;q\rangle$ and $\mathcal{H}^{(00)}\langle 2;q\rangle$ respectively.

For $\alpha \in \mathbb{R}$ the matrix pseudo-group $\mathrm{SU}_\alpha(2)$ of S. Woronowicz [93, 94] which is a deformation of $\mathcal{SU}[2]$ is obtained from $\mathcal{U}\langle 2\rangle$ by dividing by the $*$-ideal in $\mathcal{U}\langle 2\rangle$ generated by the elements

$$x_{12} - \alpha x_{21}^* \text{ and } x_{22} - x_{11}^*$$

which is a coideal of $\mathcal{U}\langle 2;1\rangle$. Moreover, $\mathrm{SU}_\alpha(2)$ is a $*$-Hopf algebra with the antipode given by $x_{kl} \mapsto x_{lk}^*$.

We treat deformations of H(2), $\mathrm{H}^{(0)}(2)$ and $\mathrm{H}^{(00)}(2)$. We have

1.8.2. PROPOSITION. *Let $\alpha, \beta \in \mathbb{C}_0$. The two-sided $*$-ideal $\mathcal{I}_{\alpha,\beta}$ in $\mathcal{H}\langle 2\rangle$ generated by the elements*

$$\begin{gathered} zx - \alpha xz, \quad z^*x - \overline{\alpha} xz^*, \\ yx - \beta xy, \quad y^*x - \overline{\beta} xy^*, \\ [z, y], \quad [z^*, y], \\ [z, z^*], \quad [y, y^*] \end{gathered}$$

is a graded coideal of $\mathcal{H}\langle 2;q\rangle$. Moreover, the antipode of $\mathcal{H}\langle 2;q\rangle$ leaves $\mathcal{I}_{\alpha,\beta}$ invariant.

PROOF: We have

$$\begin{aligned} \Delta(zx - \alpha xz) &= ((z \otimes z)(z \otimes x + x \otimes y) - \alpha(z \otimes x + x \otimes y)(z \otimes z) \\ &= z^2 \otimes (zx - \alpha xz) + (zx - \alpha xz) \otimes yz + \alpha xz \otimes (yz - zy) \end{aligned}$$

which shows that

$$\Delta(zx - \alpha xz) \subset \mathcal{I}_{\alpha,\beta} \otimes \mathcal{H}\langle 2\rangle + \mathcal{H}\langle 2\rangle \otimes \mathcal{I}_{\alpha,\beta}.$$

Next we have

$$S(zx - \alpha xz) = -z^{-1}(xz^{-1} - \alpha z^{-1}x)y^{-1}.$$

Similarly, for the other elements of $\mathcal{I}_{\alpha,\beta}$.◇

We denote the deformation $\mathcal{H}\langle 2;q\rangle/\mathcal{I}_{\alpha,\beta}$ of H(2) by $\mathrm{H}_{q;\alpha,\beta}(2)$. By the above proposition $\mathrm{H}_{q;\alpha,\beta}(2)$ is an involutive q-Hopf algebra. In a similar way, the involutive q-Hopf algebra $\mathrm{H}^{(0)}_{q;\beta}(2)$, $\beta \in \mathbb{C}_0$, and the involutive q-bialgebra $\mathrm{H}^{(00)}_{q;\beta}(2)$, $\beta \in \mathbb{C}$, are defined.

6. MATRIX $*$-BIALGEBRA OVER A FINITE GROUP.: Let $\mathbb{L}$ be finite and let $\mathcal{M}\langle\mathbb{L}\rangle$ be the free $*$-algebra generated by indeterminates $x_{\sigma,\tau}$, $\sigma, \tau \in \mathbb{L}$. The algebra $\mathcal{M}\langle\mathbb{L}\rangle$ becomes an $\mathbb{L}^{\mathrm{inv}}$-graded $*$-algebra if we put

$$\mathrm{d}(x_{\sigma,\tau}) = \sigma\tau^{-1},$$

and an $\mathbb{L}^{\mathrm{inv}}$-module $*$-algebra with

$$\sigma.x_{\tau,\varrho} = x_{\sigma\tau,\sigma\varrho}.$$

If we set

$$\Delta x_{\sigma,\tau} = \sum_{\rho \in \mathbb{L}} x_{\sigma,\varrho} \otimes x_{\varrho,\tau}$$

and

$$\varepsilon x_{\sigma,\tau} = \delta_{\sigma,\tau}$$

this extends to an involutive (γ, α)-bialgebra structure on $\mathcal{M}\langle\mathbb{L}\rangle$ which in general is not of commutation factor type. Moreover, the ideal generated by the unitarity relations (1.8.1) is an α-invariant, $\mathbb{L}$-graded $*$-coideal so that we can divide $\mathcal{M}\langle\mathbb{L}\rangle$ by this ideal to obtain another involutive (γ, α)-bialgebra. We denote it by $\mathcal{U}\langle\mathbb{L}\rangle$.

7. q-DEFORMATION OF Env(sl(2)): For $\beta \in \mathbb{R}$, $\beta \neq 0$, $\beta \neq \pm 1$, we consider the $*$-algebra generated by the indeterminates $\tilde{x}, \tilde{y}$ and $\tilde{y}^{-1}$ where $\tilde{y}$ is assumed to be hermitian and the relations

$$\tilde{y}\tilde{y}^{-1} = \tilde{y}^{-1}\tilde{y} = \mathbf{1} \tag{1.8.3}$$

$$\tilde{y}\tilde{x} = \beta\tilde{x}\tilde{y} \tag{1.8.4}$$

$$[\tilde{x}, \tilde{x}^*] = \frac{\tilde{y}^2 - \tilde{y}^{-2}}{\beta - \beta^{-1}} \tag{1.8.5}$$

are assumed to hold. If $\tilde{y}$ is defined to be group-like and $\tilde{x}$ to be $(\tilde{y}^{-1}, \tilde{y})$-primitive this is the β-deformation $\mathrm{sl}_\beta(2)$ of the universal enveloping algebra Env(sl(2)) of the Lie algebra sl(2) of SL(2); see [25]. Indeed, relation (1.8.4) and its adjoint version become $[\tilde{x}, h] = \tilde{x}$ and $[\tilde{x}^*, h] = -\tilde{x}^*$ if we, heuristically, put $y = \beta^h$, differentiate and let $q \to 1$. For $q \to 1$ relation (1.8.5) becomes $[\tilde{x}, \tilde{x}^*] = -2h$. The $*$-bialgebra $\mathrm{sl}_\beta(2)$ is a $*$-Hopf algebra with the antipode given by $\tilde{x} \mapsto -\tilde{y}^{-1}\tilde{x}\tilde{y}$ and $\tilde{x}^* \mapsto -\tilde{y}^{-2}\tilde{x}^*\tilde{y}$.

Assume that we can take the square root of the generator y of the $*$-Hopf algebra $\mathrm{H}^{(0)}_{1;\beta^2}(2)$ which amounts to adding hermitian indeterminates $y^{1/2}$ and $y^{-1/2}$ and relations $y^{1/2}y^{-1/2} = y^{-1/2}y^{1/2} = \mathbf{1}$, $(y^{1/2})^2 = y$, $y^{1/2}x = \beta x y^{1/2}$. If we put $\tilde{x} = xy^{-1/2}$ and $\tilde{y} = y^{1/2}$ we obtain $\mathrm{sl}_\beta(2)$ as the quotient of this $*$-Hopf algebra by the relation (1.8.5).

1.9. White noise on involutive bialgebras. Let $\mathcal{B}$ be an involutive (γ, α)-bialgebra. A quantum stochastic process $(j_{st})_{(s,t)\in T_{\mathbb{L}}}$ on $\mathcal{B}$ over a probability space $(\mathcal{A}, \Phi)$ is called a *white noise* on $\mathcal{B}$ if it satisfies (WN1)-(WN3) and the additional condition

(WN4) (increment property)

$$j_{rs} \star j_{st} = j_{rt} \text{ for all } 0 \leq r \leq s \leq t \leq T$$

$$j_{tt} = \epsilon \text{ id for all } t \geq 0$$

We also speak of (γ, α)-white noise. Similarly, there is (α, γ)-white noise on (α, γ)-bialgebras. In the commutation factor case we have the notions of left and right q-white noise.

Let $\mathcal{B}$ be an involutive (γ, α)-Hopf algebra. For a quantum stochastic process $(j_t)_{t\geq 0}$ on $\mathcal{B}$ we define the increments j_{st} by

$$j_{st} = (j_s \circ S) \star j_t.$$

Then j_{st} satisfy the increment property. We call the process j_t a white noise if j_{st} forms a white noise.

We have the following proposition.

1.9.1. PROPOSITION.

(i) *A family* $(F_t)_{t\in[0,L]}$ *of quantum random vectors on the double-module* $*$*-vector space* V *is an additive* (γ,α)*-white noise if and only if* $(j_{st})_{(s,t)\in T_L}$ *with* $j_{st} = \mathcal{T}(F_t - F_s)$ *form a white noise on the involutive* (γ,α)*-Hopf algebra* $\mathcal{T}(V)$.

(ii) *A family* $(A_{st})_{(s,t)\in T_L}$ *of quantum random* $d\times d$*-matrices is a multiplicative white noise if and only if* j_{st} *with* $j_{st} = \mathcal{T}(A_{st})$ *form a white noise on the involutive q-bialgebra* $\mathcal{M}\langle d\rangle$.

(iii) *A family* $(U_t)_{t\in[0,L]}$ *of unitary operators on a Hilbert space* $\mathbb{C}^d \otimes \mathcal{H}$ *with* $\mathcal{H}$ *a Hilbert space is a unitary q-white noise if and only if* $(j_{st})_{(s,t)\in T_L}$ *with* $j_{st} = \mathcal{T}(U_s^{-1}U_t)$ *form a white noise on the involutive q-bialgebra* $\mathcal{U}\langle d;q\rangle$.

PROOF: (i): We must prove the increment property. We have by construction that $j_{rs}\star j_{st} : \mathcal{T}(V) \to \mathcal{A}$ is a $*$-algebra homomorphism. The proof of (i) is complete if we show that $j_{rs}\star j_{st}$ and j_{rt} agree on the generators v of $\mathcal{T}(V)$. But this is clear, because we have

$$(j_{rs}\star j_{st})(v) = F_{rs}(v) + F_{st}(v) = F_{rt}(v) = j_{rt}(v).$$

(ii) and (iii) are proved in the same manner.◇

A linear functional ψ on an involutive q-bialgebra $\mathcal{B}$ is called *conditionally positive* if

$$\psi(b^*b) \geq 0 \text{ for all } b\in\mathcal{B} \text{ with } \epsilon(b) = 0.$$

1.9.2. THEOREM. *Let* $(j_{st})_{(s,t)\in T_L}$ *be a white noise on the involutive* (γ,α)*-bialgebra* $\mathcal{B}$*. Then the 1-dimensional distributions* φ_t *form a 1-parameter convolution semi-group of states and there exists a unique linear functional* ψ *on* $\mathcal{B}$ *such that*

$$\varphi_t = \exp_\star(t\psi). \tag{1.9.1}$$

Moreover,

$$\psi(b) = \frac{\mathrm{d}}{\mathrm{d}t}\varphi_t(b)|_{t=0} \text{ for all } b\in\mathcal{B} \tag{1.9.2}$$

and ψ *is hermitian, conditionally positive and* α*-invariant and satifies* $\psi(1) = 1$.

PROOF: We have

$$\begin{aligned}
\varphi_s\star\varphi_t &= \varphi_s\star\varphi_{s,s+t}\\
&= (\Phi\circ j_{0s})\otimes(\Phi\circ j_{s,s+t})\circ\Delta\\
&= (\Phi\otimes\Phi)\circ(j_{0s}\otimes j_{s,s+t})\circ\Delta\\
&= \Phi\circ j_{0,s+t}\\
&= \varphi_{s+t}
\end{aligned}$$

where we only used (WN1) and the factorization property (1.3.1) for time ordered correlations. We have

$$\mathcal{T}(\varphi_s)\mathcal{T}(\varphi_t) = \mathcal{T}(\varphi_{s+t})$$

and for $\mathcal{D}\in\mathcal{U}(\mathcal{B})$

$$\lim_{t\downarrow 0}\mathcal{T}(\varphi_t)\lceil\mathcal{D} = \mathrm{id}$$

which by a well known result for 1-parameter semi-groups of matrices yields the existence of a unique $R_{\mathcal{D}} \in \mathrm{L}(D)$ such that

$$\mathcal{T}(\varphi_t)\lceil\mathcal{D} = \exp(tR_{\mathcal{D}})$$

for all $t \geq 0$. Clearly, $(R_{\mathcal{D}})_{\mathcal{D}\in\mathcal{U}(\mathcal{B})}$ satisfies the consistency condition and thus

$$\varphi_t = \exp_\star(t\psi)$$

with $\psi = \epsilon \circ R$ and R the inductive limit of $(R_{\mathcal{D}})$. For $b \in \operatorname{Kern} \epsilon$ we have

$$\psi(b^*b) = \lim_{t\downarrow 0} \varphi_t(b^*b) \geq 0.$$

Since the limit of hermitian linear functionals is hermitian and $(\varphi_t - \epsilon)(1) = 0$, this means that ψ is hermitian, conditionally positive and vanishes at **1** by (1.9.2).◇

We call ψ the (infinitesimal) generator of the white noise. The choice of (γ, α) rather than (α, γ) can be justified by the following. We show that for white noise the pyramidal correlations factorize; cf. [68].

1.9.3. PROPOSITION. *Let $(j_{st})_{(s,t)\in T_{\mathrm{L}}}$ be a white noise on the involutive (γ, α)-bialgebra $\mathcal{B}$. Then*

$$\begin{aligned} &\Phi(j_{t_1t_2}(b_1)\dots j_{t_nt_{n+1}}(b_n)j_{t_nt_{n+1}}(c_n)\dots j_{t_1t_2}(c_1)) \\ &= \varphi_{t_2-t_1}(b_1c_1)\dots\varphi_{t_{n+1}-t_n}(b_nc_n) \end{aligned} \tag{1.9.3}$$

for all $n \in \mathbb{N}$, $0 < t_1 < \dots < t_{n+1}$, $b_1, \dots, b_n, c_1, \dots, c_n \in \mathcal{B}$.

PROOF: We have

$$\begin{aligned} &j_{t_1t_2}(b_1)\dots j_{t_{n-1}t_n}(b_{n-1})j_{t_nt_{n+1}}(b_nc_n)j_{t_{n-1}t_n}(c_{n-1})\dots j_{t_1t_2}(c_1) \\ &= j_{t_1t_2}(b_1)\dots j_{t_{n-1}t_n}(b_{n-1}c_{n-1}) \\ &\qquad j_{t_nt_{n+1}}((b_nc_n).\mathrm{d}(c_{n-1}))j_{t_{n-2}t_{n-1}}(b_{n-2})\dots j_{t_1t_2}(c_1) \\ &= j_{t_1t_2}(b_1c_1)j_{t_2t_3}(b_1c_1).\mathrm{d}(c_1))\dots j_{t_nt_{n+1}}((b_nc_n).\mathrm{d}(c_{n-1}\dots c_1)). \end{aligned}$$

Applying Φ to this expression and using the fact that $\varphi_{t_{k+1}-t_k}$ are α-invariant, we obtain the right hand side of (1.9.3).◇

Next we investigate the relationship between (γ, α)- and (α, γ)-white noise.

1.9.4. PROPOSITION. *Let $(j_{st})_{(s,t)\in T_{\mathrm{L}}}$, $\mathrm{L} < \infty$, be a (γ, α)-white noise. Then the time-reversed process $(j_{st}^{\mathrm{op}})_{(s,t)\in T_{\mathrm{L}}}$ with*

$$j_{st}^{\mathrm{op}} = j_{\mathrm{L}-t,\mathrm{L}-s}$$

is a white noise on the involutive (α, γ)-bialgebra $\mathcal{B}^{\mathrm{op}}$. Moreover, (j_{st}) and j_{st}^{op} have the same generator and the same 1-dimensional distributions.

PROOF: We have

$$\begin{aligned} (j_{rs}^{\mathrm{op}} \star^{\mathrm{op}} j_{st}^{\mathrm{op}})(b) &= \sum j_{\mathrm{L}-s,\mathrm{L}-r}(\mathrm{d}(b_{(1)}).b_{(2)})j_{\mathrm{L}-t,\mathrm{L}-s}(b_{(1)}) \\ &= \sum j_{\mathrm{L}-t,\mathrm{L}-s}(b_{(1)})j_{\mathrm{L}-s,\mathrm{L}-r}(b_{(2)}) \\ &= j_{rt}^{\mathrm{op}}(b) \end{aligned}$$

which is the increment property. Similarly, the independence of increments is checked. Finally, we have

$$\Phi \circ j_{st}^{op} = \varphi_{L-t,L-s} = \varphi_{t-s} = \Phi \circ j_{st}$$

which proves (WN3) and (WN4) and the last statement of the Proposition.◇

By Prop. 1.9.3 and 1.9.4 a (α, γ)-white noise will factorize for 'reversed' pyramidal orderings that is for valleys rather than mountains. Frequently, a (α, γ)-white noise will be called a 'time-reversed' white noise. We will see later that it really can happen that a white noise does not factorize on valleys, so that, in particular, the time-reversed white noise differs from the original white noise.

We show how to reconstruct a white noise from its 1-dimensional distributions, that is from its infinitesimal generator. Let $\mathcal{S}_{\geq 2}$ be the directed set of all finite subsets σ of $\mathbb{R}_+$ such that $\#\sigma \geq 2$ with the partial ordering given by the inclusion of sets. For $\sigma = \{t_1 < \cdots < t_{n+1}\} \in \mathcal{S}_{\geq 2}$ we put $\check{\alpha} = t_1$, $\hat{\alpha} = t_{n+1}$ and

$$\mathcal{B}_\sigma = \mathcal{B}_{t_1t_2} \otimes_{(\gamma,\alpha)} \mathcal{B}_{t_2t_3} \otimes_{(\gamma,\alpha)} \cdots \otimes_{(\gamma,\alpha)} \mathcal{B}_{t_nt_{n+1}}$$

where $\mathcal{B}_{st}$ for $s < t$ denotes a copy of $\mathcal{B}$. For $\sigma \subset \tau$ we define the $*$-algebra homomorphism

$$\tilde{j}_{\sigma,\tau} : \mathcal{B}_\sigma \to \mathcal{B}_\tau$$

as follows. If $\check{\beta} = \check{\alpha}$, $\hat{\beta} = \hat{\alpha}$ and $\#\sigma = 2$ we set $\tilde{j}_{\sigma,\tau} = \Delta_{\#\tau-1}$. For $\check{\beta} = \check{\alpha}$ and $\hat{\beta} = \hat{\alpha}$ we then put

$$\tilde{j}_{\sigma,\tau} = \tilde{j}_{\{t_1,t_2\},\tau(t_1,t_2)} \otimes \ldots \otimes \tilde{j}_{\{t_nt_{n+1}\},\tau(t_nt_{n+1})}$$

where for $s, t \in \sigma$ we set $\tau(s,t) = \{r \in \tau : s \leq r \leq t\}$. In the case when $\tau = \tau_1 \cup \tau' \cup \tau_2$ and $s_1 < s' < s_2$ for all $s_1 \in \tau_1$, $s' \in \tau'$, $s_2 \in \tau_2$, such that $\check{\tau}' = \check{\alpha}$ and $\hat{\tau}' = \hat{\alpha}$ we put $\tilde{j}_{\sigma,\tau}(b) = \mathbf{1} \otimes \tilde{j}_{\sigma,\tau'}(b) \otimes \mathbf{1}$. For example, if $\tau = \{t_1 < \cdots < t_6\}$ and $\sigma = \{t_1 < t_3 < t_6\}$ then $\tilde{j}_{\sigma,\tau} = \Delta_2 \otimes \Delta_3$. The situation is described in the following diagram

$$\begin{array}{ccc} (\mathcal{B}_{t_1t_2} \otimes \mathcal{B}_{t_2t_3}) & \otimes & (\mathcal{B}_{t_3t_4} \otimes \mathcal{B}_{t_4t_5} \otimes \mathcal{B}_{t_5t_6}) \\ \uparrow \Delta_2 & & \uparrow \Delta_3 \\ \mathcal{B}_{t_1t_3} & \otimes & \mathcal{B}_{t_3t_6}. \end{array}$$

Notice that this is just the dual construction of the algebraic part of the reconstruction of a semi-group-valued classical stochastic process with independent stationary multiplicative increments out of its 1-dimensional distributions. The role of the multiplication is now played by the comultiplication but in reversed direction. From the general coassociativity law

$$(\Delta_{m_1} \otimes \ldots \otimes \Delta_{m_k}) \circ \Delta_k = \Delta_{\sum m_i}$$

we obtain for $\sigma \subset \tau \subset \gamma$

$$\tilde{j}_{\tau,\gamma} \circ \tilde{j}_{\sigma,\tau} = \tilde{j}_{\sigma,\tau}.$$

Next define the state φ_σ on $\mathcal{B}_\sigma$ to be the product state $\varphi_{t_2-t_1} \otimes \ldots \otimes \varphi_{t_{n+1}-t_n}$. Then

$$\varphi_\sigma = \varphi_\tau \circ \tilde{j}_{\sigma,\tau},$$

since $\{\varphi_t\}$ is a convolution semi-group. Thus $(\mathcal{B}_\sigma, \varphi_\sigma, \tilde{j}_{\sigma,\tau})$ forms an inductive system in the category of $*$-algebras and states. Denote the inductive limit of this system by $(\tilde{\mathcal{A}}, \tilde{\Phi}, \tilde{j}_\sigma)$. We put $\tilde{j}_{st} = \tilde{j}_{\{s,t\}}$.

1.9.5. PROPOSITION. *The family $(\tilde{j}_{st})_{(s,t)\in T_L}$ forms a white noise on $\mathcal{B}$ with generator ψ.*

PROOF: If $\check{\alpha} = \hat{\beta}$ we have

$$\tilde{M} \circ (\tilde{j}_\sigma \otimes \tilde{j}_\tau) = \tilde{M} \circ (\tilde{j}_{\sigma\cup\tau} \otimes \tilde{j}_{\sigma\cup\tau}) \circ (\tilde{j}_{\sigma,\sigma\cup\tau} \otimes \tilde{j}_{\tau,\sigma\cup\tau}) = \tilde{j}_{\sigma\cup\tau}$$

because $\tilde{j}_{\sigma\cup\tau}$ is an algebra homomorphism. From this we obtain for $r < s < t$

$$\begin{aligned}\tilde{j}_{rs} \star \tilde{j}_{st} &= \tilde{M} \circ (\tilde{j}_{rs} \otimes \tilde{j}_{st}) \circ \Delta \\ &= \tilde{j}_{\{r,s,t\}} \circ \Delta \\ &= \tilde{j}_{\{r,s,t\}} \circ \tilde{j}_{\{r,t\},\{r,s,t\}} \\ &= \tilde{j}_{rt}\end{aligned}$$

which is the increment property (WN1). Next for $s_1 < t_1 < s_2 < t_2$ and $b_1, b_2 \in \mathcal{B}$

$$\begin{aligned}\tilde{j}_{s_1t_1}(b_1) j_{s_2t_2}(b_2) &= \tilde{j}_{s_1t_1}(b_1)\tilde{j}_{s_2t_2}(b_2) \\ &= \tilde{M} \circ (\tilde{j}_{s_1t_1}(b_1) \otimes j_{s_2t_2}(b_2)) \\ &= \tilde{j}_{\{s_1,t_1,s_2,t_2\}}(b_1 \otimes b_2) \\ &= \tilde{j}_{\{s_1,t_1,s_2,t_2\}}((b_1 \otimes \mathbf{1})(\mathbf{1} \otimes b_2)) \\ &= \tilde{j}_{\{s_1,t_1,s_2,t_2\}}((\mathbf{1} \otimes \mathrm{d}(b_1).b_2)(b_1 \otimes \mathbf{1})) \\ &= \tilde{j}_{s_2t_2}(\mathrm{d}(b_1).b_2)\tilde{j}_{s_1t_1}(b_1).\end{aligned}$$

We have by construction that

$$\begin{aligned}\tilde{\Phi}(\tilde{j}_{t_1t_2}(b_1) \ldots \tilde{j}_{t_nt_{n+1}}(b_n)) &= \varphi_\sigma(b_1 \otimes \ldots b_n) \\ &= \varphi_{t_2-t_1}(b_1) \ldots \varphi_{t_{n+1}-t_n}(b_n)\end{aligned}$$

which together with the fact that ψ is α-invariant yields the factorization property. Thus (WN2) is proved.◇

The white noise $(\tilde{j}_{st})$ has the following universal property.

1.9.6. PROPOSITION. *Let $(j_{st})_{(s,t)\in T_L}$ be a white noise on $\mathcal{B}$ with generator ψ. Then there exists a unique $*$-algebra homomorphism $\Theta : \tilde{\mathcal{A}} \to \mathcal{A}$ such that*

$$j_{st} = \Theta \circ \tilde{j}_{st}, \qquad \tilde{\Phi} = \Phi \circ \Theta.$$

In particular, (j_{st}) and $(\tilde{j}_{st})$ are equivalent.

PROOF: We define $j_\sigma : \mathcal{B}_\sigma \to \mathcal{A}$ by

$$j_\sigma = M_n \circ (j_{t_1t_2} \otimes \ldots j_{t_nt_{n+1}})$$

which by the commutation rules is a $*$-algebra homomorphism. The increment property yields $\tilde{j}_{\sigma,\tau} \circ j_\sigma = j_\tau$ and the factorization property gives $\Phi \circ j_\sigma = \varphi_\sigma$. By the property of an inductive limit it follows that there exists a unique $*$-algebra homomorphism Θ such that $j_\sigma = \Theta \circ \tilde{j}_\sigma$ and $\tilde{\Phi} = \Phi \circ \Theta$.$\diamondsuit$

Now we have the *reconstruction theorem.*

1.9.7. COROLLARY. *Let $\{\varphi_t : t \in \mathbb{R}_+\}$ be a 1-parameter convolution semi-group of α-invariant states on an involutive (γ, α)-bialgebra $\mathcal{B}$, and suppose that*

$$\lim_{t \downarrow 0} \varphi_t(b) = \epsilon(b) \text{ for all } b \in \mathcal{B}.$$

Then there is a white noise on $\mathcal{B}$ with 1-dimensional distributions equal to φ_t. Moreover, two white noises on $\mathcal{B}$ are equivalent if and only if their 1-dimensional distributions, and thus their generators, agree.$\diamondsuit$

In Section 2.2 we will see that any conditionally positive linear functional appears as the generator of a white noise.

2. Symmetric white noise on Bose Fock space.

This chapter is the heart of the theory. We show that an arbitrary symmetric white noise can be realized as a process of linear operators on some suitable Bose Fock space. The operators will, in general, be far from bounded, but they always will be defined on a common invariant dense domain in the underlying Fock space. Our operators are kernel operators in the sense of H. Maassen; see [51]. The original work of Maassen is generalized by considering kernels with three arguments which take values in spaces of operators on Hilbert spaces. This generalization is necessary to include *differential second quantization processes* on the one hand and Bose Fock spaces over spaces of square-integrable functions on $\mathbb{R}_+$ *taking values in an arbitrary Hilbert space* on the other hand. There has already been done a considerable amount of work in this direction by P.A. Meyer, M. Lindsay and K. R. Parthasarathy. In particular, we widely made use of [49, 50, 57]. We arranged the preparatory part of this chapter always in view of our main theorem which is the representation of white noise on Fock space.

2.1. Bose Fock space over $L^2(\mathbb{R}_+, H)$.

Let $(\mathcal{A}_1, \Phi_1), \ldots, (\mathcal{A}_n, \Phi_n)$ be quantum probability spaces. We form the product quantum probability space $(\mathcal{A}, \Phi)$ with

$$\mathcal{A} = \mathcal{A}_1 \otimes \ldots \otimes \mathcal{A}_n,$$
$$\Phi = \Phi_1 \otimes \ldots \otimes \Phi_n.$$

Then the subalgebras $\mathcal{A}_1, \ldots, \mathcal{A}_n$ of $\mathcal{A}$ are independent in the sense of Bose (or symmetric) independence, i.e. Φ factorizes, and $\mathcal{A}_k$ and $\mathcal{A}_l$ commute for $k \neq l$. On the Hilbert space level, if we have $(\mathcal{H}_1, \Omega_1), \ldots, (\mathcal{H}_n, \Omega_n)$ with Hilbert spaces $\mathcal{H}_k$ and unit vectors $\Omega_k \in \mathcal{H}_k$, the algebras $\mathbf{B}(\mathcal{H}_1), \ldots, \mathbf{B}(\mathcal{H}_n)$ are independent as subspaces of the quantum probability space $(\mathbf{B}(\mathcal{H}), \Omega)$ where $\mathcal{H} = \mathcal{H}_1 \otimes \ldots \otimes \mathcal{H}_n$ and Φ the state given by the unit vector $\Omega = \Omega_1 \otimes \ldots \otimes \Omega_n$ in $\mathcal{H}$. This shows that tensor products are the natural device to build Bose independent spaces. *Continuous* tensor products should be the right objects for natural realizations of our continuous parameter white noise processes. The functor $\mathcal{F}$ which associates with a Hilbert space K the Bose Fock space $\mathcal{F}(K)$ over K has the property that direct sums of Hilbert spaces are transformed into tensor products of Hilbert spaces. If K is a continuous sum of Hilbert spaces $\mathcal{F}(K)$ will be a continuous tensor product of Hilbert spaces. Examples of (homogeneous) sums of Hilbert spaces are the spaces $L^2(\mathbb{R}_+, H)$ of square integrable functions on $\mathbb{R}_+$ with values in a Hilbert space H. Indeed, we will see that Fock spaces over continuous sums of this form are the natural structure for white noise to be build on.

Let $\mathcal{H}$ be a Hilbert space. The Bose Fock space $\mathcal{F}(\mathcal{H})$ over $\mathcal{H}$ is the complete direct sum

$$\overline{\bigoplus_{n \in \mathbb{N}}} \mathcal{H}^{\overline{\otimes}_s n}$$

of the n-fold *symmetric* Hilbert space tensor product $\mathcal{H}^{\overline{\otimes}_s n}$ of $\mathcal{H}$ with itself where we put $\mathcal{H}^{\overline{\otimes}_s 0} = \mathbb{C}$. For $f \in \mathcal{H}$ the coherent state, or exponential vector, $\mathrm{E}(f) \in \mathcal{F}(\mathcal{H})$ is the vector

$$\mathrm{E}(f) = \bigoplus_{n \in \mathbb{N}} \frac{f^{\otimes n}}{\sqrt{n!}}.$$

It is well known that $\{\mathrm{E}(f) : f \in \mathcal{H}\}$ forms a total set of linearly independent vectors in $\mathcal{F}(\mathcal{H})$. One checks that

$$\langle \mathrm{E}(f), \mathrm{E}(g)\rangle = \mathrm{e}^{\langle f,g\rangle}.$$

For a Hilbert space H we denote by $L^2(\mathbb{R}_+, H)$ the Hilbert space of H-valued square-integrable functions on $\mathbb{R}_+$, i.e. $f \in L^2(\mathbb{R}_+, H)$ if and only if $f : \mathbb{R}_+ \to H$ is measurable and

$$\int_{\mathbb{R}_+} \|f(t)\|^2 \mathrm{d}t < \infty.$$

We give another interpretation which will be very useful for our further investigations. Let $\mathcal{S}$ be the set of finite subsets of $\mathbb{R}_+$. We have

$$\mathcal{S} = \bigcup_{n\in\mathbb{N}} \mathcal{S}_n \tag{2.1.1}$$

where

$$\mathcal{S}_n = \{\sigma \in \mathcal{S} : \#\sigma = n\}.$$

Frequently, we identify $\mathcal{S}_n, n \geq 1$, with the subset of $\mathbb{R}_+^n$ formed by strictly ordered n-tuples through

$$\{t_1 < \cdots < t_n\} \mapsto (t_1, \ldots, t_n).$$

If we define $\mu\lceil\mathcal{S}_0$ to be the measure with mass 1 at $\emptyset$ and if we put $\mu\lceil\mathcal{S}_n$, $n \geq 1$, equal to Lebesgue measure on $\mathbb{R}_+^n$ restricted to $\mathcal{S}_n \subset \mathbb{R}_+^n$ we obtain a σ-finite measure μ on $\mathcal{S}$. We put $\mathrm{d}\mu(\omega) = \mathrm{d}\omega$. For $\omega \in \mathcal{S}$, $\omega = \{t_1 < \cdots < t_n\}$, $n \geq 1$, we put

$$H^{\overline{\otimes}\omega} = H_{t_1}\overline{\otimes} \ldots \overline{\otimes} H_{t_n}$$

where each H_{t_k} is a copy of H and we set $H^{\overline{\otimes}\emptyset} = \mathbb{C}$.

2.1.1. PROPOSITION. *The equations*

$$\mathcal{U}(\mathrm{E}(f))(\emptyset) = 1$$
$$\mathcal{U}(\mathrm{E}(f))(\omega) = f(t_1) \otimes \ldots \otimes f(t_n),$$

$\omega \subset \mathcal{S}$, $\omega = \{t_1 < \cdots < t_n\}$, $n \geq 1$, *define a unitary operator from* $\mathcal{F}(L^2(\mathbb{R}_+, H))$ *to the 'direct integral'*

$$\int_{\mathcal{S}} H^{\overline{\otimes}\omega} \mathrm{d}\omega = \overline{\bigoplus_{n\in\mathbb{N}} L^2(\mathcal{S}_n, H^{\overline{\otimes}n}, \mu\lceil\mathcal{S}_n)}$$

of the Hilbert spaces $H^{\overline{\otimes}\omega}$.

PROOF: We have for $F, G \in \int_{\mathcal{S}} H^{\overline{\otimes}\omega}\mathrm{d}\omega$

$$\begin{aligned}\langle F, G\rangle &= \int_{\mathcal{S}} \langle F(\omega), G(\omega)\rangle \mathrm{d}\omega \\ &= \overline{F(\emptyset)}G(\emptyset) + \sum_{n\in\mathbb{N}} \int_{\mathcal{S}_n} \langle F(\omega), G(\omega)\rangle \mathrm{d}\omega \\ &= \overline{F(\emptyset)}G(\emptyset) + \sum_{n\in\mathbb{N}} \int \cdots \int_{t_1<\cdots<t_n} \langle F(t_1,\ldots,t_n), G(t_1,\ldots,t_n)\rangle \mathrm{d}t_n \ldots \mathrm{d}t_n\end{aligned}$$

and for $f, g \in L^2(\mathbb{R}_+, H)$

$$\begin{aligned}&\int \cdots \int_{t_1<\cdots<t_n} \langle f(t_1)\otimes\ldots\otimes f(t_n), g(t_1)\otimes\ldots\otimes g(t_n)\rangle \mathrm{d}t_n \ldots \mathrm{d}t_1 \\ &= \int \cdots \int_{t_1<\cdots<t_n} \langle f(t_1), g(t_1)\rangle \ldots \langle f(t_n), g(t_n)\rangle \mathrm{d}t_n \ldots \mathrm{d}t_1 \\ &= \frac{1}{n!}\left(\int_{\mathbb{R}_+} \langle f(t), g(t)\rangle \mathrm{d}t\right)^n \\ &= \frac{\langle f, g\rangle^n}{n!}\end{aligned}$$

which proves that $\mathcal{U}$ is well defined on $\mathcal{F}(L^2(\mathbb{R}_+, H))$ and $\mathcal{U}$ is an isometry. It remains to show that $\mathcal{F}(L^2(\mathbb{R}_+, H))$ is the whole of $\int_{\mathcal{S}} H^{\overline{\otimes}\omega}\mathrm{d}\omega$. Clearly, the linear subspace $L^2(\mathcal{S}_n, \mathbb{C}) \otimes H^{\otimes n}$ of $L^2(\mathcal{S}_n, H^{\overline{\otimes} n})$ is dense. Moreover, the characteristic functions $\chi_{I_1\times\cdots\times I_n}$ with $I_1 \times \cdots \times I_n$ a bounded cylinder set in $\mathcal{S}_n$ are total in $L^2(\mathcal{S}_n, \mathbb{C})$. Since $\mathcal{S}_n$ consists of strictly ordered tuples we must have $I_1 < \cdots < I_n$. Thus

$$\chi_{I_1\times\cdots\times I_n}(t_1,\ldots,t_n)\otimes\xi_1\otimes\ldots\xi_n = f(t_1)\otimes\ldots\otimes f(t_n)$$

if we define $f \in L^2(\mathbb{R}_+, H)$ to be the function taking the value ξ_k on I_k and vanishing elsewhere. Then $\mathcal{U}$ maps $\mathrm{E}(f)$ to

$$\chi_{I_1\times\cdots\times I_n}\otimes\xi_1\otimes\ldots\otimes\xi_n \in L^2(\mathbb{R}_+, \mathbb{C})\otimes H^{\otimes\omega}.$$

This completes the proof.◇

It is possible to give a rigorous definition of $\int_{\mathcal{S}} H^{\overline{\otimes}\omega}\mathrm{d}\omega$ as a direct integral of Hilbert spaces in the sense of [24].

The following proposition plays an important role in most of the computations to come; cf. [49, 50].

2.1.2.PROPOSITION (INTEGRAL-SUM IDENTITY). *For $n, k \in \mathbb{N}$, $k \geq 1$, and*

$$F^{(n)} : \mathcal{S}^n \to \mathbb{C}^k$$

integrable with respect to $\mu^{\otimes n}$ the function

$$F : \mathcal{S} \to \mathbb{C}^k$$

with

$$F(\omega) = \sum_{\sigma_1 \dot\cup \ldots \dot\cup \sigma_n = \omega} F^{(n)}(\sigma_1, \ldots, \sigma_n)$$

is μ-integrable and

$$\int_S F(\omega)\mathrm{d}\omega = \int_{S^n} F^{(n)}(\sigma_1, \ldots, \sigma_n)\mathrm{d}\sigma_1 \ldots \mathrm{d}\sigma_n. \qquad (2.1.2)$$

PROOF: It is sufficient to treat the case $n = 2$, the other cases follow by induction on n. We have

$$\begin{aligned}
&\int_S \Big(\sum_{\sigma \cup \tau = \omega} F^{(2)}(\sigma, \tau)\Big)\mathrm{d}\omega \\
&= F^{(2)}(\emptyset, \emptyset) + \sum_{n=1}^{\infty} \frac{1}{n!} \int_{\mathbb{R}_+^n} \Big(\sum_{\sigma \cup \tau = \{t_1, \ldots, t_n\}} F^{(2)}(\sigma, \tau)\Big)\mathrm{d}t_1 \ldots \mathrm{d}t_n \\
&= F^{(2)}(\emptyset, \emptyset) + \sum_{n=1}^{\infty} \sum_{\pi \in \mathrm{S}_n} \sum_{k=0}^{n} \frac{1}{n!} \frac{1}{k!(n-k)!} \\
&\qquad \Big(\int_{\mathbb{R}_+^n} F^{(2)}(\{t_{\pi(1)}, \ldots, t_{\pi(k)}\}, \{t_{\pi(k+1)}, \ldots, t_{\pi(n)}\})\mathrm{d}t_1 \ldots \mathrm{d}t_n\Big) \\
&= F^{(2)}(\emptyset, \emptyset) + \sum_{\substack{k,l \in \mathbb{N} \\ k+l \neq 0}} \Big(\int_{S_k} \int_{S_l} F^{(2)}(\sigma, \tau)\mathrm{d}\sigma\mathrm{d}\tau\Big)
\end{aligned}$$

which is (2.1.2) for $n = 2$.$\Diamond$

If $D \subset H$ is a linear subspace we denote by $\mathcal{P}_0(D)$ the set of finite-dimensional linear subspaces of D and by

$$\mathcal{F}(L^2(\mathbb{R}_+, D)) = \int_S D^{\otimes \omega}\mathrm{d}\omega$$

the linear subspace of $\mathcal{F}(L^2(\mathbb{R}_+, H))$ consisting of all elements which come from measurable vector fields F with $F(\omega) \in D^{\otimes \omega}$ for almost all $\omega \in \mathcal{S}$. Let N denote the number operator on $\mathcal{F}(L^2(\mathbb{R}_+, H))$. For $\alpha \in \mathbb{R}_+$ the operator α^N is defined by the functional calculus for the self-adjoint operator N. We have

$$\mathrm{dom}(\alpha^N) = \{F \in \mathcal{F}(L^2(\mathbb{R}_+, H)) : \sum_{n \in \mathbb{N}} \alpha^{2n} \|F^{(n)}\|^2 < \infty\}.$$

We form the linear subspace

$$\mathcal{E}_D = \Big(\bigcup_{E \in \mathcal{P}_0(D)} \int_S E^{\otimes \omega}\mathrm{d}\omega\Big) \cap \Big(\bigcap_{\alpha \in \mathbb{R}_+} \mathrm{dom}(\alpha^N)\Big)$$

of $\mathcal{F}(L^2(\mathbb{R}_+, D))$ which is dense in $\mathcal{F}(L^2(\mathbb{R}_+, H))$ if D is dense in H. In other words, $\mathcal{E}_D$ consists of all $F \in \mathcal{F}(L^2(\mathbb{R}_+, H))$ with the properties

(E1) there is a finite-dimensional linear subspace E of D such that $F(\omega) \in E^{\otimes\omega}$ for almost all $\omega \in \mathcal{S}$

(E2) the integral $\int_{\mathcal{S}} \alpha^{\omega} \|F(\omega)\|^2 d\omega$ is finite for all $\alpha \in \mathbb{R}_+$.

2.2. Kernels and operators. Let D be a fixed dense linear subspace of the Hilbert space H. We introduce a class of linear operators on $\mathcal{E}_D$. The operators are given by *kernels.* A kernel is a mapping

$$k : \mathcal{S}^3 \to \dot{\bigcup_{\omega_1,\omega_2 \in \mathcal{S}}} \mathbf{L}(D^{\otimes\omega_1}, D^{\otimes\omega_2})$$

such that

$$k(\sigma, \tau, \varrho) \in \mathbf{L}(D^{\otimes(\tau\cup\varrho)}, D^{\otimes(\sigma\cup\tau)})$$

and

$$k(\sigma, \tau, \varrho) = 0 \text{ unless } \sigma, \tau, \varrho \text{ are disjoint.}$$

Our kernels will be in the class $\mathcal{K}_D$ consisting of all kernels with the properties

(K1) for all $E_1 \in \mathcal{P}_0(D)$ there is a $E_2 \in \mathcal{P}_0(D)$ such that

$$k(\sigma, \tau, \varrho) E_1^{\otimes(\tau\cup\varrho)} \subset E_2^{\otimes(\sigma\cup\tau)}$$

for all $\sigma, \tau, \varrho \in \mathcal{S}$

(K2) for all $E_1 \in \mathcal{P}_0(D)$ and E_2 as in (K1) the mapping

$$\mathcal{S}_{n_1} \dot{\times} \mathcal{S}_{n_2} \dot{\times} \mathcal{S}_{n_3} \to \mathbf{L}(E_1^{\otimes(n_2+n_3)}, E_2^{\otimes(n_1+n_2)})$$

with

$$(\sigma, \tau, \varrho) \mapsto k(\sigma, \tau, \varrho) \lceil E_1^{\otimes(n_2+n_3)}$$

is measurable for all choices of $n_1, n_2, n_3 \in \mathbb{N}$

(K3) for all $E \in \mathcal{P}_0(D)$ there is a constant $c \in \mathbb{R}_+$ such that

$$\|k(\sigma, \tau, \varrho) \lceil E^{\otimes(\tau\cup\varrho)}\| \leq c^{\sigma\cup\tau\cup\varrho}$$

for all $\sigma, \tau, \varrho \in \mathcal{S}$

(K4) there is a bounded subset C of $\mathbb{R}_+$ such that

$$k(\sigma, \tau, \varrho) = 0 \text{ unless } \sigma \cup \tau \cup \varrho \subset C.$$

For $\sigma, \tau \in \mathcal{S}$ the pre-Hilbert space $D^{\otimes(\sigma\cup\tau)}$ is isomorphic to $D^{\otimes\sigma} \otimes D^{\otimes\tau}$ in a natural way. We map

$$\overrightarrow{\bigotimes_{t \in \sigma\cup\tau}} \xi_t \mapsto (\overrightarrow{\bigotimes_{t\in\sigma}} \xi_t) \otimes (\overrightarrow{\bigotimes_{t\in\tau}} \xi_t)$$

where for $\omega = \{t_1 < \cdots < t_n\}$, $n \geq 1$,

$$\overrightarrow{\bigotimes_{t\in\omega}} \xi_t = \xi_{t_1} \otimes \ldots \otimes \xi_{t_n}.$$

Denote this isomorphism by $\iota_{\sigma,\tau}$. Now we define the *ampliation* $k^\varepsilon(\sigma,\tau,\varrho)$ of $k(\sigma,\tau,\varrho)$ to $\mathbf{L}(D^{\otimes(\varepsilon\cup\tau\cup\varrho)}, D^{\otimes(\varepsilon\cup\sigma\cup\tau)})$, $\sigma,\tau,\varrho,\varepsilon$ pairwise disjoint, by

$$k^\varepsilon(\sigma,\tau,\varrho) = (\iota_{\varepsilon,\sigma\cup\tau})^{-1} \circ (\mathrm{id} \otimes k(\sigma,\tau,\varrho)) \circ \iota_{\varepsilon,\tau\cup\varrho}.$$

In the sequel, we drop the index ε in $k^\varepsilon(\sigma,\tau,\varrho)$ since it will always be clear to which space $k(\sigma,\tau,\varrho)$ is to be applied.

2.2.1. LEMMA. *For $F \in \mathcal{F}(L^2(\mathbb{R}_+, E))$, $E \in \mathcal{P}_0$, and $k \in \mathcal{K}_D$ the estimate*

$$\int_S \alpha^\omega \Big(\sum_{\sigma\dot\cup\tau\dot\cup\varepsilon=\omega} \int_S \|k(\sigma,\tau,\varrho)F(\varepsilon\cup\tau\cup\varrho)\| \mathrm{d}\rho\Big)^2 \mathrm{d}\omega$$
$$\leq \mathrm{e}^{\mu(C)9\alpha c^2} \int_S (9\alpha c^2)^\omega \|F(\omega)\|^2 \mathrm{d}\omega$$

holds for all $\alpha \in \mathbb{R}_+$.

PROOF: Using the inequality

$$\Big(\sum_{i=1}^n \alpha_i\Big)^2 \leq n \sum_{i=1}^n \alpha_i^2$$

which holds for all $n \in \mathbb{N}$ and all $\alpha_1, \ldots, \alpha_n \in \mathbb{R}$, we have that the left hand side is less than

$$\int_S \alpha^\omega 3^\omega \sum_{\sigma\dot\cup\tau\dot\cup\varepsilon=\omega} \Big(\int_S \|k(\sigma,\tau,\varrho)F(\varepsilon\cup\tau\cup\varrho)\| \mathrm{d}\rho\Big)^2 \mathrm{d}\omega$$
$$= \int_{S^3} (3\alpha)^{\sigma\cup\tau\cup\varepsilon} \Big(\int_S \|k(\sigma,\tau,\varrho)F(\varepsilon\cup\tau\cup\varrho)\| \mathrm{d}\rho\Big)^2 \mathrm{d}\sigma \mathrm{d}\tau \mathrm{d}\varepsilon$$

where we used the integral-sum identity. Using the Cauchy-Schwartz inequality and (K3) and (K4), this is less than ($c \geq 1$ without loss of generality)

$$\mathrm{e}^{\mu(C)9\alpha c^2} \int_{S^3} (3\alpha c^2)^\tau (3\alpha)^\varepsilon \|F(\varepsilon\cup\tau\cup\varrho)\|^2 \mathrm{d}\varrho \mathrm{d}\varepsilon \mathrm{d}\tau$$

which, again by the integral-sum identity, is smaller than

$$\mathrm{e}^{\mu(C)9\alpha c^2} \int_S (3\alpha c^2)^\omega \Big(\sum_{\varepsilon\dot\cup\tau\dot\cup\varrho=\omega} \|F(\omega)\|^2\Big) \mathrm{d}\omega$$

which is equal to the right hand side.◇

The above lemma establishes the existence of

$$\omega \mapsto \sum_{\sigma \dot\cup \tau \dot\cup \varepsilon = \omega} \int_{\mathcal{S}} k(\sigma,\tau,\varrho)F(\varepsilon \cup \tau \cup \varrho)\mathrm{d}\varrho$$

as an element of $\mathcal{E}_D$ if F is an element of $\mathcal{E}_D$. We denote this function by $(\underline{k}F)(\omega)$ and obtain a linear operator $\underline{k}$ on $\mathcal{E}_D$.

The estimates in the proof of Lemma 3.2.1 show that a wider class of kernels could have been admitted. It is possible to replace (K4) and (K5) by the weaker condition

for all $E \in \mathcal{P}_0(D)$ there is a constant $c \in \mathbb{R}_+$ and a measurable function $l : \mathcal{S}^2 \to \mathbb{C}$ with

$$\int_{\mathcal{S}^2} \alpha^{\sigma \cup \varrho} |l(\sigma,\varrho)|^2 \mathrm{d}\sigma \mathrm{d}\varrho < \infty$$

for all $\alpha \in \mathbb{R}_+$ such that

$$\|k(\sigma,\tau,\varrho)\lceil E^{\otimes(\tau\cup\varrho)}\| \le c^{\tau} l(\sigma,\varrho)$$

for all $\sigma,\tau,\varrho \in \mathcal{S}$.

This is a mixture of a 2-dimensional version of our conditions on the elements of $\mathcal{E}_D$ and of a condition (K3) for the variable τ only. The condition was established in [49] for the case $H = \mathbb{C}$. However, for our purposes the smaller class $\mathcal{K}_D$ is sufficient.

The following proposition gives a criterion for convergence of operators coming from kernels in $\mathcal{K}_D$.

2.2.2. Proposition. *Let k_n, $n \in \mathbb{N}$, and k be in $\mathcal{K}_D$ and suppose that the quantities E_2, c and C appearing in (K1)-(K4) can be chosen the same for k_n, $n \in \mathbb{N}$, and k. Then*

$$\lim_{n\to\infty} k_n(\sigma,\tau,\varrho)\lceil E^{\otimes(\tau\cup\varrho)} = k(\sigma,\tau,\varrho)\lceil E^{(\tau\cup\varrho)}$$

for all $\sigma,\tau,\varrho \in \mathcal{S}$ and $E \in \mathcal{P}_0(D)$ implies that $\underline{k}_n$ converges strongly on $\mathcal{E}_D$ to $\underline{k}$.

Proof: We may assume without loss of generality that $k(\sigma,\tau,\varrho) = 0$. Clearly, for $E_1 \in \mathcal{P}_0(D)$ and $F \in \int_{\mathcal{S}} E_1^{\otimes\omega}\mathrm{d}\omega$ we have

$$\lim_{n\to\infty} \sum_{\sigma \dot\cup \tau \dot\cup \varepsilon = \omega} k_n(\sigma,\tau,\varrho)F(\tau \cup \varrho \cup \varepsilon) = 0.$$

Moreover, the estimate

$$\begin{aligned} &\|\sum_{\sigma \dot\cup \tau \dot\cup \varepsilon = \omega} k_n(\sigma,\tau,\varrho)F(\tau\cup\varrho\cup\varepsilon)\| \\ &\le \sum_{\sigma \dot\cup \tau \dot\cup \varepsilon = \omega} c^{\sigma\cup\tau\cup\varrho}\chi_{\mathcal{S}\cap C}(\sigma\cup\varrho)\|F(\tau\cup\varrho\cup\varepsilon)\| \end{aligned}$$

holds for all $\omega, \varrho \in \mathcal{S}$. The right hand side of this inequality is an integrable function of ϱ for all $\omega \in \mathcal{S}$ and by Lebesgue's dominated convergence theorem

$$\lim_{n\to\infty} \|(\underline{k}_n F)(\omega)\| = 0$$

for all ω. Next $\|(\underline{k}_n F)(\omega)\|^2$ is dominated by

$$\sum_{\sigma\dot{\cup}\tau\dot{\cup}\varepsilon=\omega} 3^{\omega} e^{c^2\mu(C)}(c^2)^{\sigma\cup\tau}\chi_{S\cap C}(\sigma)\int_S \|F(\tau\cup\varrho\cup\varepsilon)\|^2 d\rho,$$

which by the integral-sum identity and by property (E2) of F is an integrable function of ω. Again by Lebesgue's theorem we conclude that

$$\lim_{n\to\infty}\int_S \|(\underline{k}_n F)(\omega)\|^2 d\omega = 0$$

which proves the proposition.◇

The mapping from $\mathcal{K}_D$ to $\mathbf{L}(\mathcal{E}_D)$ which sends k to $\underline{k}$ is linear. In a sense, it is also injective.

2.2.3. THEOREM. *For $k\in\mathcal{K}_D$ we have that $\underline{k}\in\mathbf{L}(\mathcal{E}_D)$ is equal to 0 if and only if for each $E\in\mathcal{P}_0(D)$*

$$k(\sigma,\tau,\varrho)\lceil E^{\otimes(\tau\cup\varrho)} = 0$$

for almost all $(\sigma,\tau,\varrho)\in\mathcal{S}^3$.

PROOF: Let $E_1\in\mathcal{P}_0(D)$ be fixed. Then there is an $E_2\in\mathcal{P}_0(D)$ such that $\underline{k}(\mathcal{E}_{E_1})\subset\mathcal{E}_{E_2}$ by property (K2). By the integral-sum identity we have for $F\in\mathcal{E}_{E_1}$ and $G\in\mathcal{F}(H)$

$$\begin{aligned}\langle\underline{k}F,G\rangle &= \int_S \langle(\underline{k}F)(\omega),G(\omega)\rangle d\omega\\ &= \int_{S^4}\langle k(\sigma,\tau,\varrho)F(\tau\cup\varrho\cup\varepsilon),G(\sigma\cup\tau\cup\varepsilon)\rangle d\varrho d\sigma d\tau d\varepsilon\end{aligned}$$

which shows that $\underline{k}\lceil\mathcal{E}_{E_1} = 0$ if $k(\sigma,\tau,\varrho)\lceil E_1^{\otimes(\tau\cup\varrho)}$ vanishes a.e. Now let $\underline{k}=0$. We introduce the transform k' of k by setting

$$k'(\sigma,\tau,\varrho) = \sum_{\varepsilon\subset\tau} k(\sigma,\varepsilon,\varrho).$$

It is easy to see that k' is again in $\mathcal{K}_D$ if k is, and that k can be recovered from k' via a Möbius inversion with respect to τ, i.e.

$$k(\sigma,\tau,\varrho) = \sum_{\varepsilon\subset\tau}(-1)^{\tau\setminus\varepsilon}k'(\sigma,\tau,\varrho);$$

see [49]. Moreover, we can express $(\underline{k}F)(\omega)$ by the formula

$$(\underline{k}F)(\omega) = \sum_{\sigma\subset\omega}\int_S k'(\sigma,\sigma^c,\varrho)F(\sigma^c\cup\varrho)d\varrho \tag{2.2.1}$$

where σ^c denotes $\omega\setminus\sigma$. But $\underline{k}F=0$ means that

$$\int_S \|(\underline{k}F)(\omega)\|^2 d\omega = 0$$

and thus $(\underline{k}F)(\omega) = 0$ a.e. We wish to write the right hand side of (2.2.1) as an integral only, and, to that goal, we introduce the following auxiliary Hilbert space $\mathcal{F}_\omega$. For a fixed $\omega \in \mathcal{S}$ let λ_ω denote the measure on $\mathbb{R}_+$ that equals Lebesgue measure on $\mathbb{R}_+ \setminus \omega$ but has mass 1 on each of the points of ω. Denote by μ_ω the measure on $\mathcal{S}$ whose restriction to $\mathcal{S}_n$ equals the restriction of the product measure $(\lambda_\omega)^{\otimes n}$ on $\mathbb{R}_+^n$ to $\mathcal{S}_n$ and which has mass 1 at $\emptyset$. Then $\mathcal{F}_\omega$ is the direct integral

$$\int_{\mathcal{S}} H^{\otimes \varepsilon} \mu(\mathrm{d}\varepsilon) = \overline{\bigoplus_{n \in \mathbb{N}}} L^2(\mathcal{S}_n, H^{\overline{\otimes} n}, \mu_\omega \lceil \mathcal{S}_n).$$

For a μ_ω-integrable function $\Phi : \mathcal{S} \to \mathbb{C}$ we have

$$\int_{\mathcal{S}} \Phi(\varepsilon)\mu_\omega(\mathrm{d}\varepsilon) = \sum_{\sigma \subset \omega} \int_{\mathcal{S}} \Phi(\sigma \cup \varrho)\mathrm{d}\varrho$$

and F is in $\mathcal{F}_\omega$ if and only if

$$\int_{\mathcal{S}} \|F(\varepsilon)\|^2 \mu_\omega(\mathrm{d}\varepsilon) = \sum_{\sigma \subset \omega} \int_{\mathcal{S}} \|F(\sigma \cup \varrho)\|^2 \mathrm{d}\varrho < \infty,$$

that is if and only if $\varrho \mapsto F(\sigma \cup \varrho)$ is in $\mathcal{F}(L^2(\mathbb{R}_+, H))$ for all $\sigma \subset \omega$. Equation (2.2.1) becomes

$$(\underline{k}F)(\omega) = \int_{\mathcal{S}} k'_\omega(\varepsilon)F(\varepsilon)\mu_\omega(\mathrm{d}\varepsilon)$$

for all $F \in \mathcal{E}_D$ with

$$k'_\omega(\varepsilon) = k'((\varepsilon \cap \omega)^c, \varepsilon \cap \omega, \varepsilon \setminus \omega) \in \mathbf{L}(E_1^{\otimes \varepsilon}, E_2^{\otimes \omega}).$$

Then $k'_\omega \lceil \mathcal{S}_n$ is square-integrable with respect to $(\lambda_\omega)^{\otimes n}$. Since a $(\lambda_\omega)^{\otimes n}$-square-integrable function is also square-integrable with respect to $\lambda^{\otimes n}$ we have

$$\int_{\mathcal{S}_n} k'_\omega(\varepsilon)F(\varepsilon)(\lambda_\omega)^{\otimes n}(\mathrm{d}\varepsilon) = 0$$

for all $F \in L^2(\mathcal{S}_n, E_1^{\otimes n}, (\lambda_\omega)^{\otimes n})$. In general, if $(X, \mathcal{E}, \nu)$ is a measure space and

$$A : X \to \mathbf{L}(\mathbb{C}^{n_2}, \mathbb{C}^{n_1})$$

is ν-square-integrable then

$$\int_X A(x)\Psi(x)\nu(\mathrm{d}x) = 0$$

for all $\Psi \in L^2(X, \mathbb{C}^{d_2}, \nu)$ clearly implies that $A(x) = 0$ a.e. If we apply this to $X = \mathcal{S}_n$, $n_1 = (\dim E_1)^n$ and $n_2 = (\dim E_2)^\omega$ we obtain that for almost all $\varrho \in \mathcal{S}$

$$k'(\sigma, \sigma^c, \varrho)\lceil E_1^{\otimes(\sigma^c \cup \varrho)} = 0$$

for all $\sigma \subset \omega$. But this gives that

$$k'(\sigma, \tau, \varrho)\lceil E_1^{\otimes(\tau \cup \varrho)} = 0$$

for almost all $(\sigma,\tau,\varrho)\in\mathcal{S}^3$. Finally, we have

$$\begin{aligned}\sum_{\mathcal{S}^3}\|k(\sigma,\tau,\varrho)\|\mathrm{d}\sigma\mathrm{d}\tau\mathrm{d}\varrho &\le \frac{1}{2}\int_{\mathcal{S}^3}\sum_{\alpha\dot\cup\beta=\tau}(\|k'(\sigma,\alpha,\varrho)\|+\|k'(\sigma,\beta,\varrho)\|)\mathrm{d}\sigma\mathrm{d}\tau\mathrm{d}\varrho\\ &= \frac{1}{2}\int_{\mathcal{S}^4}(\|k'(\sigma,\alpha,\varrho)\|+\|k'(\sigma,\beta,\varrho)\|)\mathrm{d}\sigma\mathrm{d}\alpha\mathrm{d}\varrho\mathrm{d}\beta\\ &= 0\end{aligned}$$

(where we omitted the $\lceil E_1^{\otimes\cdots}$). This means that

$$k(\sigma,\tau,\varrho)\lceil E_1^{\otimes(\tau\cup\varrho)}=0$$

for almost all $(\sigma,\tau,\varrho)\in\mathcal{S}^3$.$\diamondsuit$

The next theorem shows that the set of operators on $\mathcal{E}_D$ which come from kernels in $\mathcal{K}_D$ is closed under composition.

2.2.4. THEOREM. *For $k_1,k_2\in\mathcal{K}_D$ we define the kernel k_1k_2 by*

$$\begin{aligned}&(k_1k_2)(\sigma,\tau,\varrho)\lceil E^{\otimes(\tau\cup\varrho)}\\ &=\sum_{\substack{\sigma_1\dot\cup\sigma_2\dot\cup\sigma_3=\sigma\\ \tau_1\dot\cup\tau_2\dot\cup\tau_3=\tau\\ \varrho_1\dot\cup\varrho_2\dot\cup\varrho_3=\varrho}}\int_{\mathcal{S}}k_1(\sigma_1,\sigma_2\cup\tau_1\cup\tau_2,\varrho_1\cup\varrho_2\cup\varepsilon)\times\\ &\qquad\times k_2(\varepsilon\cup\sigma_2\cup\sigma_3,\tau_2\cup\tau_3\cup\varrho_2,\varrho_3)\lceil E^{\otimes(\tau\cup\varrho)}\mathrm{d}\varepsilon\end{aligned}$$

for σ,τ,ϱ disjoint and $E\in\mathcal{P}_0(D)$. Then k_1k_2 is again in $\mathcal{K}_D$ and $\underline{k_1k_2}=\underline{k}_1\circ\underline{k}_2$.

PROOF: The existence of the integral follows from

$$\begin{aligned}\int_{\mathcal{S}^{10}}&\|k_1(\sigma_1,\sigma_2\cup\tau_1\cup\tau_2,\varrho_1\cup\varrho_2\cup\varepsilon)\times\\ &\times k_2(\varepsilon\cup\sigma_2\cup\sigma_3,\tau_2\cup\tau_3\cup\varrho_2,\varrho_3)\|\mathrm{d}\varepsilon\mathrm{d}\sigma_1\ldots\mathrm{d}\varrho_3<\infty\end{aligned}$$

which holds by property (K3) and (K4) for k_1 and k_2, and by another application of the integral-sum identity. This also shows that (K1), (K2) and (K3) are satisfied by k_1k_2. Clearly, (K4) is fulfilled with $C=C_1\cup C_2$ and we have $k_1k_2\in\mathcal{K}_D$. Moreover, for $F \in \mathcal{E}_D$ and $\omega \in \mathcal{S}$, applying the integral-sum identity twice, we obtain

$$
\begin{aligned}
&((\underline{k_1} \circ \underline{k_2})F)(\omega) \\
&= \sum_{\sigma \dot\cup \tau \dot\cup \varepsilon = \omega} \int_{\mathcal{S}} k_1(\sigma, \tau, \varrho)\Big(\sum_{\sigma' \dot\cup \tau' \dot\cup \varepsilon' = \varepsilon \cup \tau \cup \varrho} \int_{\mathcal{S}} k_2(\sigma', \tau', \varrho') F(\varepsilon' \cup \tau' \cup \varrho') \mathrm{d}\varrho' \Big) \mathrm{d}\varrho \\
&= \sum_{\substack{\sigma \dot\cup \sigma_1' \dot\cup \tau_1' \dot\cup \varepsilon_1' \\ \dot\cup \sigma_2' \dot\cup \tau_2' \dot\cup \varepsilon_2' = \omega}} \int_{\mathcal{S}^2} \Big(\sum_{\sigma_3' \dot\cup \tau_3' \dot\cup \varepsilon_3' = \varrho} k_1(\sigma, \sigma_2' \cup \tau_2' \varepsilon_2', \varrho) k_2(\sigma_1' \cup \sigma_2' \cup \sigma_3', \tau_1' \cup \tau_2' \cup \tau_3', \varrho') \times \\
&\qquad\qquad \times F(\varepsilon_1' \cup \varepsilon_2' \cup \varepsilon_3' \cup \tau_1' \cup \tau_2' \cup \tau_3' \cup \varrho') \Big) \mathrm{d}\varrho \mathrm{d}\varrho' \\
&= \sum_{\substack{\sigma \dot\cup \sigma_1' \dot\cup \tau_1' \dot\cup \varepsilon_1' \\ \dot\cup \sigma_2' \dot\cup \tau_2' \dot\cup \varepsilon_2' = \omega}} \int_{\mathcal{S}^4} k_1(\sigma, \sigma_2' \cup \tau_2' \cup \varepsilon_2', \sigma_3' \cup \tau_3' \cup \varepsilon_3') k_2(\sigma_1' \cup \sigma_2' \cup \sigma_3', \tau_1' \cup \tau_2' \tau_3', \varrho') \times \\
&\qquad\qquad \times F(\varepsilon_1' \cup \varepsilon_2' \cup \varepsilon_3' \cup \tau_1' \cup \tau_2' \cup \tau_3' \cup \varrho') \mathrm{d}\sigma_3' \mathrm{d}\tau_3' \mathrm{d}\varepsilon_3' \mathrm{d}\varrho' \\
&= \sum_{\alpha \dot\cup \beta \dot\cup \varepsilon_1' = \omega} \int_{\mathcal{S}} \sum_{\substack{\sigma \dot\cup \sigma_1' \dot\cup \sigma_2' = \alpha \\ \tau_2' \dot\cup \varepsilon_2' \dot\cup \tau_1' = \beta \\ \tau_3' \dot\cup \varepsilon_3' \dot\cup \varrho' = \gamma}} \Big(\int_{\mathcal{S}} k_1(\sigma, \sigma_2' \cup \tau_2' \cup \varepsilon_2', \sigma_3' \cup \tau_3' \cup \varrho_3') \times \\
&\qquad\qquad \times k_2(\sigma_1' \cup \sigma_2' \cup \sigma_3', \tau_1' \cup \tau_2' \cup \tau_3', \varrho') \mathrm{d}\sigma_3' \Big) \\
&\qquad\qquad \times F(\beta \cup \gamma \cup \varepsilon_1') \mathrm{d}\gamma \\
&= \sum_{\alpha \dot\cup \beta \dot\cup \varepsilon_1' = \omega} \int_{\mathcal{S}} (k_1 k_2)(\alpha, \beta, \gamma) F(\beta \cup \gamma \cup \varepsilon_1') \mathrm{d}\gamma \\
&= ((\underline{k_1 k_2})F)(\omega). \diamondsuit
\end{aligned}
$$

Next we tackle the question under which conditions on k the operator $\underline{k}$ is in $\mathbf{H}(\mathcal{E}_D)$.

2.2.5. Proposition. *Let $k \in \mathcal{K}_D$ such that*

$$k(\sigma, \tau, \varrho) \in \mathbf{H}(D^{\otimes(\tau \cup \varrho)}, D^{\otimes(\sigma \cup \tau)}) \tag{2.2.2}$$

for all $\sigma, \tau, \varrho \in \mathcal{S}$ and such that the kernel $\tilde{k}$ with

$$\tilde{k}(\sigma, \tau, \varrho) = k(\varrho, \tau, \sigma)^*$$

satisfies condition (K1). Then $\tilde{k} \in \mathcal{K}_D$, $\underline{k} \in \mathbf{H}(\mathcal{E}_D)$ and $(\underline{k})^* = \underline{\tilde{k}}$.

Proof: Clearly, $\tilde{k}$ satisfies conditions (K2)-(K4) so that $\tilde{k} \in \mathcal{K}_D$. We have for $F, G \in \mathcal{E}_D$

$$
\begin{aligned}
\langle \underline{k} F, G \rangle &= \int_{\mathcal{S}^4} \langle k(\sigma, \tau, \varrho) F(\tau \cup \varrho \cup \varepsilon), G(\sigma \cup \tau \cup \varepsilon) \rangle \mathrm{d}\sigma \mathrm{d}\tau \mathrm{d}\varepsilon \mathrm{d}\varrho \\
&= \int_{\mathcal{S}^4} \langle F(\tau \cup \varrho \cup \varepsilon), k(\sigma, \tau, \varrho)^* G(\sigma \cup \tau \cup \varepsilon) \rangle \mathrm{d}\sigma \mathrm{d}\tau \mathrm{d}\varepsilon \mathrm{d}\varrho \\
&= \langle F, \underline{\tilde{k}} G \rangle. \diamondsuit
\end{aligned}
$$

We denote by $\mathcal{K}_{D,h}$ the subset of $\mathcal{K}_D$ consisting of those k that satisfy (2.2.2) and which have the property that $\tilde{k} \in \mathcal{K}_D$.

2.3. The basic formula. In this section we state the formula for a realization on Bose Fock space of a given symmetric white noise on a $*$-bialgebra. The proof will be given in Section 2.5 with the help of quantum stochastic calculus which we develop in Section 2.4.

Let D be a dense linear subspace of a Hilbert space H and let $\xi \in D$ and $T \in \mathbf{L}(D)$. For $t \in \mathbb{R}_+$ define the linear operators

$$\begin{aligned} a^*(t)(\xi) &: \mathbb{C} \to D^{\otimes\{t\}}, \\ \lambda(t)(T) &: D^{\otimes\{t\}} \to D^{\otimes\{t\}}, \\ a(t)(\xi) &: D^{\otimes\{t\}} \to \mathbb{C} \end{aligned}$$

by

$$\begin{aligned} a^*(t)(\xi)(z) &= z\xi_t, \\ \lambda(t)(T)(\zeta_t) &= T\zeta_t, \\ a(t)(\xi)(\zeta_t) &= \langle \xi, \zeta_t \rangle \end{aligned}$$

for $z \in \mathbb{C}$, $\zeta_t \in D^{\otimes\{t\}}$ where ξ_t denotes the image of ξ under the identification of D with $D^{\otimes\{t\}}$. As before, we denote the ampliations

$$\begin{aligned} a^*(t)(\xi) &: D^{\otimes\omega} \to D^{\otimes\omega\cup\{t\}}, \\ \lambda(t)(T) &: D^{\otimes\omega\cup\{t\}} \to D^{\otimes\omega\cup\{t\}}, \\ a(t)(\xi) &: D^{\otimes\omega\cup\{t\}} \to D^{\otimes\omega} \end{aligned}$$

which exist for all $\omega \in \mathcal{S}$, $t \notin \omega$, by the same symbols. We have $a(t)(\xi)^* = a^*(t)(\xi)$ and $\lambda(t)(T)^* = \lambda(t)(T^*)$ if $T \in \mathbf{H}(D)$.

Let $\mathcal{C}$ be a coalgebra and let there be given

- a linear mapping $\eta : \mathcal{C} \to D$
- a linear mapping $\rho : \mathcal{C} \to \mathbf{L}(D)$
- an antilinear mapping $\vartheta : \mathcal{C} \to D$
- a linear functional ψ on $\mathcal{C}$.

In the sequel, the symbols η, ρ, ϑ, and ψ will always stand for mappings of the above kind. For disjoint σ, τ, ϱ, $t \in \sigma \cup \tau \cup \varrho$ and $b \in \mathcal{C}$ we introduce the symbol

$$\kappa(t;\sigma,\tau,\varrho)(b) = \begin{cases} a^*(t)(\eta(b)) & \text{if } t \in \sigma \\ \lambda(t)(\rho(b) - \epsilon(b)\mathrm{id}) & \text{if } t \in \tau \\ a(t)(\vartheta(b)) & \text{if } t \in \varrho. \end{cases}$$

2.3.1. PROPOSITION. *For $0 \le s \le t$ and $b \in \mathcal{C}$ we define the kernel $k_{st}^{(\eta,\rho,\vartheta,\psi)}(b)$ as follows. First we put $k_{st}^{(\eta,\rho,\vartheta,\psi)}(b)(\sigma,\tau,\varrho) = 0$ unless $\sigma \cup \tau \cup \varrho \subset [s,t)$. Next we put*

$k_{st}^{(\eta,\rho,\vartheta,\psi)}(b)(\emptyset,\emptyset,\emptyset) = \varphi_{t-s}(b)$ *where* $\varphi_t = \exp_\star(t\psi)$. *Finally, if* $\sigma \cup \tau \cup \varrho \subset [s,t)$ *and* $\sigma \cup \tau \cup \varrho \neq \emptyset$, *we set*

$$k_{st}^{(\eta,\rho,\vartheta,\psi)}(b)(\sigma,\tau,\varrho) = (\varphi_{t_1-s} \star \kappa(t_1;\sigma,\tau,\varrho) \star \varphi_{t_2-t_1} \star \kappa(t_2;\sigma,\tau,\varrho) \star \varphi_{t_3-t_2} \star \dots$$
$$\dots \star \varphi_{t_n-t_{n-1}} \star \kappa(t_n;\sigma,\tau,\varrho) \star \varphi_{t-t_n})(b)$$

where $\sigma \cup \tau \cup \varrho = \{t_1 < \cdots < t_n\}$.

Then $k_{st}^{(\eta,\rho,\vartheta,\psi)}(b)$ *is in* $\mathcal{K}_D$. *Moreover, for* $E_1 \in \mathcal{P}_0(D)$

$$k_{st}^{(\eta,\rho,\vartheta,\psi)}(b)(\sigma,\tau,\varrho)(E_1^{\otimes(\tau\cup\varrho)}) \subset E_2^{\otimes(\sigma\cup\tau)} \tag{2.3.1}$$

with

$$E_2 = \mathrm{Lin}\{\eta(c), (\rho(c) - \epsilon(c)\mathrm{id})\xi : c \in \mathcal{C}_b, \xi \in E_1\} \tag{2.3.2}$$

where $\mathcal{C}_b$ *denotes the smallest sub-coalgebra of* $\mathcal{C}$ *containing* b, *and*

$$\|k_{st}^{(\eta,\rho,\vartheta,\psi)}(b)(\sigma,\tau,\varrho)\lceil E_1^{\otimes(\tau\cup\varrho)}\| \leq c^{\sigma\cup\tau\cup\varrho} \tag{2.3.3}$$

where $c \in \mathbb{R}_+$ *is a constant depending on* b, E_1 *and* t *only.*

PROOF: It is easy to see that $k_{st}(b)$ (we drop the $(\eta,\rho,\vartheta,\psi)$) is indeed a kernel. It remains to show that $k_{st}(b)$ satisfies conditions (K1)-(K4). But (K4) is obvious with $C = [s,t)$. Proof of (K1)-(K3): Let $b = b_1,\dots,b_d$ be a vector space basis of $\mathcal{C}_b$ which is finite-dimensional by the fundamental theorem on coalgebras. Then we have

$$\Delta_n(b_l) = \sum_{k_1,\dots,k_n=1}^{d} c_{k_1,\dots,k_n}^{(l)} b_{k_1} \otimes \dots \otimes b_{k_n}$$

for some complex numbers $c_{k_1,\dots,k_n}^{(l)}$. Now let $\sigma,\tau,\varrho \subset [s,t)$ be disjoint and $\sigma\cup\tau\cup\varrho$ non-empty, and let $m = \#(\sigma\cup\tau\cup\varrho)$. We have

$$k_{st}(b)(\sigma,\tau,\varrho) = \sum_{k_1,\dots,k_{2m+1}=1}^{d} (c_{k_1,\dots,k_{2m+1}}^{(1)} \varphi_{t_1-s}(b_{k_1})\kappa(t_1;\sigma,\tau,\varrho)(b_{k_2})\varphi_{t_2-t_1}(b_{k_3})\dots$$
$$\dots\varphi_{t_m-t_{m-1}}(b_{k_{2m-1}})\kappa(t_m;\sigma,\tau,\varrho)(b_{k_{2m}})\varphi_{t-t_m}(b_{k_{2m+1}}))$$

and if we apply $k_{st}(b)(\sigma,\tau,\varrho)$ to some vector in $E_1^{\otimes(\tau\cup\varrho)}$, $E_1 \in \mathcal{P}_0(D)$, the result will be in $E_2^{\otimes(\sigma\cup\tau)}$ with E_2 given by (2.3.2). The vector space E_2 is finite-dimensional and does not depend on σ,τ,ϱ (and even not on s,t). Moreover, if (σ,τ,ϱ) varies through $\mathcal{S}_{n_1}\dot{\times}\mathcal{S}_{n_2}\dot{\times}\mathcal{S}_{n_3}$, $n_1,n_2,n_3 \in \mathbb{N}$, the operator $\kappa(t_l;\sigma,\tau,\varrho)(b_{k_{2l}})$ only assumes a finite number of different values. Since $\varphi_t(b)$ is continuous in t, $k_{st}(b)(\sigma,\tau,\varrho)$ is continuous in (σ,τ,ϱ) but for a finite number of jumps.

Therefore, the measurability condition (K1) is also satisfied. Next we have that

$$\begin{aligned}
&\|k_{st}(b)(\sigma,\tau,\varrho)\lceil E_1^{\otimes(\tau\cup\varrho)}\| \\
&\le d^{2m+1}\max\{\epsilon(c_{k_1,\dots,k_{2m+1}}\epsilon(: k_1,\dots,k_{2m+1}=1,\dots,d\} \\
&\quad(\max\{\varphi_r(b_{k_l}); r\in[s,t), l=1,\dots,2m+1\})^{m+1} \\
&\quad(\max\{\|\eta(b_{k_l})\|, \|\vartheta(b_{k_l})\|, \|(\rho(b_{k_l})-\epsilon(b_{k_l})\mathrm{id})\lceil E_1\| : l=1,\dots,2m+1\})^m.
\end{aligned}$$

First notice that the right-hand side of this estimate depends on m but not on the particular choice of σ,τ,ϱ with $\#(\sigma\cup\tau\cup\varrho)=m$. We take a closer look at the constants $c^{(1)}_{k_1,\dots,k_n}$. By the definition of Δ_n we have that

$$c^{(1)}_{k_1,\dots,k_n} = \sum_{u_1,\dots,u_{n-2}=1}^{d} c^{(1)}_{k_1,u_1}c^{(u_1)}_{k_2,u_2}\cdots c^{(u_{n-2})}_{k_{n-1},k_n}$$

for $n\ge 3$ which gives

$$|c^{(1)}_{k_1,\dots,k_n}| \le d^{n-2}(\max\{\epsilon(c^{(l)}_{k,u}\epsilon(: k,l,u=1,\dots,d\})^{n-1}$$

and

$$\max\{\epsilon(c_{k_1,\dots,k_{2m+1}}\epsilon(\} \le d^{2m-1}(\max\{c^{(l)}_{k,u}\})^{2m}.$$

This completes the proof that $k_{st}(b)$ satisfies (K3). We also proved (2.3.1) and (2.3.3).$\diamondsuit$

2.3.2. PROPOSITION.

(i)

$$\underline{k}_{rs}\star\underline{k}_{st} = \underline{k}_{rt},$$
$$\underline{k}_{tt} = \epsilon\ \mathrm{id}$$

for all $0\le r\le s\le t$.

(ii)

$$[\underline{k}_{st}(b),\underline{k}_{s't'}(b')] = 0$$

for all $0\le s\le t\le s'\le t'$ *and all* $b,b'\in\mathcal{C}$.

(iii)

$$\langle\Omega,\underline{k}_{st}(b)\Omega\rangle = \varphi_{t-s}(b)$$

for all $b\in\mathcal{C}$ *where* Ω *denotes the 'vacuum'* E(0).

(iv) *If* $\mathcal{C}$ *is a* $*$*-coalgebra and* $\rho(\mathcal{C})\subset\mathbf{H}(D)$, *and* ρ *and* ψ *are hermitian we put* $\vartheta=\tilde\eta$ *with* $\tilde\eta(b)=\eta(b^*)$. *Then* $k_{st}(b)$ *is in* $\mathcal{K}_{D,h}$ *and*

$$(\underline{k}_{st}(b))^* = \underline{k}_{st}(b^*),$$

i.e. $b\mapsto\underline{k}_{st}(b)$ *is a hermitian mapping from* $\mathcal{C}$ *to* $\mathbf{H}(\mathcal{E}_D)$.

PROOF: (i): We must show that $k_{rs} \star k_{st} = k_{rt}$. But

$$(k_{rs} \star k_{st})(b)(\sigma, \tau, \varrho) = \sum_i k_{rs}(b_{1i})(\sigma \cap [r,s), \tau \cap [r,s), \varrho \cap [r,s))$$
$$k_{st}(b_{2i})(\sigma \cap [s,t), \tau \cap [s,t), \varrho \cap [s,t))$$
$$= k_{rs}(b)(\sigma, \tau, \varrho)$$

by the definition of k_{rs}. Next (ii) follows from

$$k_{st}(b)k_{s't'}(b') = k_{s't'}(b')k_{st}(b).$$

(iii): We have

$$(\underline{k}_{st}(b)\Omega)(\omega) = k_{st}(b)(\omega, \emptyset, \emptyset)$$

and thus

$$\langle \Omega, \underline{k}_{st}(b)\Omega \rangle = k_{st}(b)(\emptyset, \emptyset, \emptyset) = \varphi_{t-s}(b).$$

(iv): This follows from $\Delta_n(b^*) = (\Delta_n(b))^*$, $\kappa(t; \sigma, \tau, \varrho)(b)^* = \kappa(t; \varrho, \tau, \sigma)(b^*)$ and the fact that $\kappa(t; \sigma, \tau, \varrho)(b)$ commutes with $\kappa(t'; \sigma, \tau, \varrho)(b')$ for $t, t' \in \sigma \cup \tau \cup \varrho$, $t \neq t'$, and for all $b, b' \in \mathcal{C}$.$\diamondsuit$

Let $\mathcal{A}$ be an algebra, let ϵ be an algebra homomorphism from $\mathcal{A}$ to $\mathbb{C}$, and let ρ be a representation of $\mathcal{A}$ on the vector space V. A linear mapping $\eta : \mathcal{A} \to V$ is called a (ρ, ϵ)-1-cocycle if for all $a, b \in \mathcal{A}$

$$\eta(ab) = \rho(a)\eta(b) + \eta(a)\epsilon(b).$$

Let V be a pre-Hilbert space and assume that $\rho(\mathcal{A}) \subset \mathbf{H}(V)$. An antilinear mapping $\vartheta : \mathcal{A} \to V$ is called a (ϵ, ρ)-1-cocycle if for all $a, b \in \mathcal{A}$

$$\vartheta(ab) = \overline{\epsilon(a)}\vartheta(b) + \rho(b)^*\vartheta(a).$$

For a linear functional ψ on $\mathcal{A}$ the (ϵ, ϵ)-coboundary of ψ is the bilinear form $\partial\psi$ on $\mathcal{A}$ given by

$$\partial\psi(a,b) = \epsilon(a)\psi(b) - \psi(ab) + \psi(a)\epsilon(b).$$

A proof of the following will be given in Section 2.5.

2.3.3. THEOREM. *Let $\mathcal{B}$ be a bialgebra and assume that $\rho(\mathcal{B}) \subset \mathbf{H}(D)$, and that*

- *ρ is a representation of $\mathcal{B}$*
- *η is a (ρ, ϵ)-1-cocycle*
- *ϑ is a (ϵ, ρ)-1-cocycle*
- $\partial\psi(a,b) = -\langle \vartheta(a), \eta(b) \rangle$.

Then $\underline{k}_{st}$ is a representation of $\mathcal{B}$ on $\mathcal{E}_D$ for all $(s,t) \in T$.

We now come to a construction which is central for our theory. Let $\mathcal{A}$ be a $*$-algebra and let ϵ be a $*$-algebra homomorphism from $\mathcal{A}$ to $\mathbb{C}$. Moreover, let ψ be a hermitian, conditionally positive linear functional on $\mathcal{A}$ and assume that $\psi(1) = 0$. We form the positive sesquilinear form

$$(a,b)_\psi = \psi((a - \epsilon(a)1)^*(b - \epsilon(b)1))$$

on $\mathcal{A}$ and divide by the null-space

$$\mathcal{N}_\psi = \{a \in \mathcal{A} : (a,a)_\psi = 0\}$$

of this form to obtain a pre-Hilbert space $D_\psi = \mathcal{A}/\mathcal{N}_\psi$. Denote by $\eta_\psi : \mathcal{A} \to D_\psi$ the canonical mapping. Then the following theorem holds.

2.3.4. THEOREM. *The equation*

$$\rho_\psi(a)\eta_\psi(b) = \eta_\psi(ab) - \eta_\psi(a)\epsilon(b) \tag{2.3.4}$$

defines a $*$-representation of $\mathcal{A}$ on D_ψ. Moreover, η_ψ is a (ρ,ϵ)-1-cocycle and

$$\partial\psi(a,b) = -\langle \eta_\psi(a^*), \eta_\psi(b)\rangle.$$

PROOF: We prove that (2.3.4) defines a $*$-representation. We have

$$\begin{aligned}\|\rho_\psi(a)\eta_\psi(b)\| &= \|\eta_\psi(a(b-\epsilon(b)\mathbf{1}))\| \\ &= \psi((b-\epsilon(b)\mathbf{1})^* a^* a(b-\epsilon(b)\mathbf{1})) \\ &= \langle \eta_\psi(b), \eta_\psi(a^* a(b-\epsilon(b)\mathbf{1}))\rangle \\ &\leq \|\eta_\psi(b)\|\|\eta_\psi(a^* a(b-\epsilon(b)\mathbf{1}))\|\end{aligned}$$

by the Cauchy-Schwartz inequality which shows that $\rho_\psi(a)$ is well defined. Since $a(b-\epsilon(b)\mathbf{1}) \in \operatorname{Kern}\epsilon$ it follows that ρ_ψ is a representation of $\mathcal{B}$ on D_ψ. Moreover,

$$\begin{aligned}\langle \rho_\psi(a)\eta_\psi(b), \eta_\psi(c)\rangle &= \psi((b-\epsilon(b)\mathbf{1})^* a^*(c-\epsilon(c)\mathbf{1})) \\ &= \langle \eta_\psi(b), \rho_\psi(a^*)\eta_\psi(c)\rangle\end{aligned}$$

which shows that ρ_ψ is a $*$-mapping from $\mathcal{A}$ to $\mathbf{H}(D_\psi)$.◇

Now we can state the

2.3.5. REPRESENTATION THEOREM FOR SYMMETRIC WHITE NOISE. *Let $\mathcal{B}$ be a $*$-bialgebra and assume that*

- *ρ is a $*$-representation of $\mathcal{B}$ on D*
- *η is a (ρ,ϵ)-1-cocycle*
- $\partial\psi(a,b) = -\langle \eta(a^*), \eta(b)\rangle$.

If we put $\vartheta = \tilde{\eta}$, then, in the vacuum state, $(\underline{k}_{st})_{(s,t)\in T}$ form a symmetric white noise on $\mathcal{B}$ with generator ψ. Conversely, a symmetric white noise on $\mathcal{B}$ with generator ψ is equivalent to $(\underline{k}_{st})$ with $\rho = \rho_\psi$ and $\eta = \eta_\psi$.

PROOF: We have

$$\tilde{\eta}(ab) = \eta(b^* a^*) = \rho(b)^*\tilde{\eta}(a) + \tilde{\eta}(b)\overline{\epsilon(a)}$$

which shows that ϑ is a (ϵ,ρ)-1-cocycle. Combining Prop. 2.3.2 (iv) and Theorem 2.3.3 we obtain that $\underline{k}_{st}$ is a $*$-homomorphism. Prop. 2.3.2 (i) is the increment property and 2.3.2 (ii) gives the independence of increments in the vacuum state. The stationarity of increments and the weak continuity follow from Prop. 2.3.2 (iii). Thus $\underline{k}_{st}$ form a Bose white noise on $\mathcal{B}$ and, again by 2.3.2 (iii), we know that ψ is its generator. The remaining part follows from Theorem 2.3.4.◇

2.4. Quantum stochastic integrals and quantum Ito's formula. The aim is to establish a quantum stochastic integration for 'non-anticipating' operator-valued processes on Fock space. But what are the integrators? Led by the realization of Brownian motion as a sum of annihilation and creation operator processes, two quantum stochastic Ito integrals, against the 'separated' processes of annihilation and creation operators, were introduced by R.L. Hudson and K.R. Parthasarathy [41]. However, apart from the usual Riemann-Bochner integral for operator processes, also another type of integrators was used in [41], namely processes coming from the differential second quantization procedure. It will become clear later that, for our purposes, we actually need all four types of integrators. Our theory itself will provide a justification of the choice of integrators. They will turn out to be exactly the components of an arbitrary additive white noise; see Section 4.2.

Following H. Maassen, we introduce the integrals for the kernels first. Consider the creation process

$$A_t^*(\xi) = A^*(\chi_{[0,t)}\xi)$$

for a fixed element $\xi \in D$. It will leave $\mathcal{E}_D$ invariant and we have

$$A_t(\xi)^* = A_t^*(\xi)$$

where $A_t(\xi)$ is the annihilation process on $\mathcal{E}_D$. The corresponding kernel-valued processes are

$$a_t^*(\xi)(\sigma,\tau,\varrho) = \begin{cases} a^*(t)(\xi) & \text{if } \tau = \varrho = \emptyset, \sigma = \{t\} \\ 0 & \text{otherwise} \end{cases}$$

and

$$a_t(\xi)(\sigma,\tau,\varrho) = \begin{cases} a(t)(\xi) & \text{if } \sigma = \tau = \emptyset, \varrho = \{t\} \\ 0 & \text{otherwise} \end{cases}$$

respectively, i.e. $\underline{a}_t^*(\xi) = A_t^*(\xi)$ and $\underline{a}_t(\xi) = A_t(\xi)$. In view of the basic formula for $k_{st}^{(\eta,\rho,\vartheta,\psi)}(b)$ the definition of the third kernel-valued process

$$\lambda_t(T)(\sigma,\tau,\varrho) = \begin{cases} \lambda(t)(T) & \text{if } \sigma = \varrho = \emptyset, \tau = \{t\} \\ 0 & \text{otherwise} \end{cases}$$

for $T \in \mathbf{L}(D)$ seems only natural. We set $\Lambda_t(T) = \underline{\lambda}_t(T)$. If $T \in \mathbf{H}(D)$ we have $\Lambda_t(T) \in \mathbf{H}(\mathcal{E}_D)$ and $\Lambda_t(T)^* = \Lambda_t(T^*)$. It is clear that $\Lambda_t(T)$ is nothing but the differential second quantization of the operator $f \mapsto \lambda_{[0,t)}Tf$ on $L^2(\mathbb{R}_+, H)$ where $\chi_{[0,t)}$ is identified with the corresponding multiplication operator on $L^2(\mathbb{R}_+, H)$ (it cuts f from time t onwards) and Tf is to be understood pointwise. The additive increments of all the three types of processes are adapted or non-anticipating in the following sense. For a measurable subset B of $\mathbb{R}_+$ the trace $B \cap \mathcal{S}$ of B in $\mathcal{S}$ is the subset $\{B \cap \omega : \omega \in \mathcal{S}\}$ of $\mathcal{S}$. The mapping

$$\mathcal{U}_B : \left(\int_{B\cap\mathcal{S}} H^{\overline{\otimes}\omega} d\omega\right) \otimes \left(\int_{B^c\cap\mathcal{S}} H^{\overline{\otimes}\omega} d\omega\right) \to \int_{\mathcal{S}} H^{\overline{\otimes}\omega} d\omega$$

given by

$$\mathcal{U}_B(F_1 \otimes F_2)(\omega) = \iota_{B\cap\omega, B^c\cap\omega}(F_1(\omega \cap B) \otimes F_2(\omega \cap B^c))$$

is unitary which follows if we look at exponential vectors. We identify

$$\mathcal{F}(L^2(B,H))\overline{\otimes}\mathcal{F}(L^2(B^c,H))$$

and $\mathcal{F}(L^2(\mathbb{R}_+,H))$ via $\mathcal{U}_B$. In particular, for $\mathcal{F}_s^t = \mathcal{F}(L^2([s,t),H))$ we have

$$\mathcal{F} = \mathcal{F}_0^t\overline{\otimes}\mathcal{F}_s^t\overline{\otimes}\mathcal{F}_t^\infty$$

to which corresponds a decomposition

$$(\mathcal{E}_D)_0^s \otimes (\mathcal{E}_D)_s^t \otimes (\mathcal{E}_D)_t^\infty$$

of $\mathcal{E}_D$. A process $(s,t) \mapsto \underline{k}_{st} \in \mathbf{L}(\mathcal{E}_D)$ is called *adapted* if there is a process

$$(s,t) \mapsto \underline{k}_{st}^{(0)} \in \mathbf{L}((\mathcal{E}_D)_s^t)$$

such that $\underline{k}_{st}$ is the ampliation $\mathrm{id} \otimes \underline{k}_{st}^{(0)} \otimes \mathrm{id}$ of $\underline{k}_{st}^{(0)}$. On the level of the kernels this simply means that

$$k_{st}(\sigma,\tau,\varrho) = 0 \text{ a.e. unless } \sigma \cup \tau \cup \varrho \subset [s,t).$$

We denote by $\mathcal{A}_D$ the class of all families $(k_{st})_{(s,t)\in T}$ of kernels in $\mathcal{K}_D$ such that
(A1) (k_{st}) is adapted
(A2) the vector space E_2 of (K1) can be chosen the same for all k_{st} if t stays in some bounded subset of $\mathbb{R}_+$, and the measurability condition (K2) holds if we also vary (s,t) in $T \subset \mathbb{R}_+ \times \mathbb{R}_+$
(A3) for all k_{st} one can choose the same constant c of (K3) if t stays in some bounded subset of $\mathbb{R}_+$.

By Prop. 2.3.1 the $k_{st}(b)$ associated with a triplet $(\eta,\rho,\vartheta,\psi)$ on a coalgebra $\mathcal{C}$ are examples of processes in $\mathcal{A}_D$. The set $\mathcal{A}_D$ becomes a vector space with the obvious structure. Moreover, the product $(k_{st}^{(1)}k_{st}^{(2)})_{st}$ of two elements in $\mathcal{A}_D$ is again in $\mathcal{A}_D$, and $\mathcal{A}_D$ becomes an algebra. The set $\mathcal{A}_{D,h}$ of elements (k_{st}) in $\mathcal{A}_D$ with the additional property that $k_{st} \in \mathcal{K}_{D,h}$ for all $(s,t)\in T$ is a $*$-algebra with $((k_{st}))^* = ((k_{st})^*)$.

As a motivation for the definition of the quantum stochastic integral of $(k_{st}) \in \mathcal{A}_D$ against $\mathrm{d}A_t^*$, $\mathrm{d}\Lambda_t$ and $\mathrm{d}A_t$ consider a simple process

$$k_{st}(\sigma,\tau,\varrho) = \begin{cases} k_j(\sigma,\tau,\varrho) & \text{if } r_j \le t < r_{j+1} \text{ and } \sigma\cup\tau\cup\varrho \subset [s,t) \\ 0 & \text{otherwise} \end{cases} \tag{2.4.1}$$

in $\mathcal{A}_D$ where $r_0 = s < r_1 < \cdots < r_{n+1} = t$ and where $k_j(\sigma,\tau,\varrho) \in \mathcal{K}_D$. Then the following holds for the corresponding 'Ito sums'

$$\begin{aligned}&(\sum_{j=0}^{n} k_{sr_j}(a^*_{r_{j+1}}(\xi) - a^*_{r_j}(\xi)))(\sigma,\tau,\varrho)\\ &= \sum_{j=0}^{n} k_j(\sigma\cap[s,r_j),\tau\cap[s,r_j),\varrho\cap[s,r_j))\times\\ &\qquad \times a^*_{r_j,r_{j+1}}(\xi)(\sigma\cap[r_j,r_{j+1}),\tau\cap[r_j,r_{j+1}),\varrho\cap[r_j,r_{j+1}))\end{aligned}$$

where we put $a^*_{r,s} = a^*_s - a^*_r$. We assume that the partition $r_0 < \cdots < r_{n+1}$ of $[s,t)$ is finer than the one given by $\sigma \cup \tau \cup \varrho \cap [s,t)$ which we can do because for any given partition $r_0 < \cdots < r_{n+1}$ we have a representation of $k_{st}(\sigma,\tau,\varrho)$ in the form (2.4.1). Then the above sum is equal to 0 unless $\sigma \neq \emptyset$, $\sigma = \{t_1 < \cdots < t_n\}$, $s \le t_1$, $s \le r < t_n < t$ for all $r \in \tau \cup \varrho$. In this case, it is equal to

$$k_j(\{t_1 < \cdots < t_{n-1}\}, \tau, \varrho) a(t_n)(\xi)$$

where j is such that $r_j \le t < r_{j+1}$. In other words, the Ito sums converge to

$$k_{st_n}(\{t_1 < \cdots < t_{n-1}\}, \tau, \varrho) a(t_n)(\xi)$$

if $\sigma \neq \emptyset$, $t_n < t$, and to 0 otherwise. Similarly, for the integrators $\lambda_t(T)$ and $a_t(\xi)$. In the sequel, if $\omega \neq \emptyset$, we denote by $\hat\omega$ the maximum of ω and we set $\overline{\omega} = \omega \setminus \{\hat\omega\}$. The integrals of $(k_{st}) \in \mathcal{A}_D$ with respect to $a^*_t(\xi)$, $\lambda_t(T)$ and $a_t(\xi)$, $\xi \in D$, $T \in \mathbf{L}(D)$, are defined by

$$\Big(\int_s^t k_{sr}\, \mathrm{d}a^*_r(\xi)\Big)(\sigma,\tau,\varrho) = \begin{cases} k_{s\hat\sigma}(\overline{\sigma},\tau,\varrho) a^*(\hat\sigma)(\xi) & \text{if } \sigma \neq \emptyset \text{ and } s \le \hat\sigma < t \\ 0 & \text{otherwise,} \end{cases}$$

$$\Big(\int_s^t k_{sr}\, \mathrm{d}\lambda_r(T)\Big)(\sigma,\tau,\varrho) = \begin{cases} k_{s\hat\tau}(\sigma,\overline{\tau},\varrho) \lambda(\hat\tau)(T) & \text{if } \tau \neq 0 \text{ and } s \le \hat\tau < t \\ 0 & \text{otherwise,} \end{cases}$$

and

$$\Big(\int_s^t k_{sr}\, \mathrm{d}a_r(\xi)\Big)(\sigma,\tau,\varrho) = \begin{cases} k_{s\hat\varrho}(\sigma,\tau,\overline{\varrho}) a(\hat\varrho)(\xi) & \text{if } \rho \neq \emptyset \text{ and } s \le \hat\rho < t \\ 0 & \text{otherwise.} \end{cases}$$

Notice that for (σ,τ,ϱ) and s fixed all the three types of integrals are constant equal to 0 as a function of t until $t = \hat\sigma$ (or $= \hat\tau$ or $= \hat\varrho$). Then there is a jump and the integrals are again constant for $t > \hat\sigma$ $(\hat\tau, \hat\varrho)$.

Since for fixed (σ,τ,ϱ) and $E \in \mathcal{P}_0(D)$ the function

$$r \mapsto k_{sr}(\sigma,\tau,\varrho) \lceil E^{\otimes(\tau\cup\varrho)}$$

is locally bounded and measurable, we can define

$$\Big(\int_s^t k_{sr}\mathrm{d}r\Big)(\sigma,\tau,\varrho) \lceil E^{\otimes(\tau\cup\varrho)}$$

to be

$$\int_s^t (k_{sr}(\sigma,\tau,\varrho) \lceil E^{\otimes(\tau\cup\varrho)})\mathrm{d}r$$

which defines the kernel $\int_s^t k_{sr} \times \mathrm{d}r$. We use the symbol $\mathrm{d}m_{\nu,r}$ to denote the differentials $\mathrm{d}a^*_r(\xi)$, $\mathrm{d}_r(T)$, $\mathrm{d}a_r(\xi)$, $\mathrm{d}r$ for $\nu = 1,2,3,4$ respectively.

2.4.1. PROPOSITION. *For* $\nu = 1,2,3,4$ *the integrals*

$$(k_{st}) \mapsto (\int_s^t k_{s\tau}\mathrm{d}m_{\nu,\tau})$$

are linear operators on $\mathcal{A}_D$ *and on* $\mathcal{A}_{D,h}$ *and for* $(k_{st}) \in \mathcal{A}_{D,h}$

$$(\int_s^t k_{s\tau}\mathrm{d}m_{\nu,\tau})^* = \int_s^\tau k_{s\tau}^*\mathrm{d}m_{\nu,\tau}^*.$$

PROOF: Clearly, the integrals are linear, and it is straightforward to check that $(\int_s^t k_{s\tau}\mathrm{d}m_{\nu,\tau})$ is again in $\mathcal{A}_D$. For $(k_{st}) \in \mathcal{A}_{D,h}$ we have for $\sigma \neq \emptyset$

$$(k_{s\hat{\sigma}}(\overline{\sigma},\tau,\varrho)a^*(\hat{\sigma})(\xi))^* = k_{s\hat{\sigma}}(\overline{\sigma},\tau,\varrho)^*a(\hat{\sigma})(\xi) = k_{s\hat{\sigma}}^*(\varrho,\tau,\overline{\sigma})a(\hat{\sigma})(\xi)$$

which proves that

$$(\int_s^t k_{s\tau}\mathrm{d}a^*(\xi))^* = \int_s^t k_{s\tau}^*\mathrm{d}a(\xi).$$

Similarly, for the other cases.◇

The following simple lemma will be useful.

2.4.2. LEMMA. *Let* $(f_{st})_{(s,t)\in T}$ *be a family of mappings*

$$f_{st} : \mathcal{S} \to \mathbb{C}$$

such that $f_{st}(\omega) = 0$ *unless* $\omega \subset [s,t)$ *and such that the functions*

$$(r,\omega) \mapsto f_{s\tau}(\omega)$$

from $[s,t) \times \mathcal{S}$ *to* $\mathbb{C}$ *are integrable with respect to* $\lambda \otimes \mu$ *for each* $s \in \mathbb{R}_+$. *Then*

$$\int_s^t (\int_{\mathcal{S}} f_{s\tau}(\omega)\mathrm{d}\omega)\mathrm{d}r = \int_{\mathcal{S}\cap[s,t)\setminus\mathcal{S}_0} f_{s\hat{\sigma}}(\overline{\sigma})\mathrm{d}\sigma.$$

PROOF: The left hand side equals

$$\int_s^t (f_{s\tau}(\emptyset) + \sum_{n=1}^\infty (\int\cdots\int_{0<t_1<\cdots<t_n} f_{s\tau}(\{t_1,\ldots,t_n\})\mathrm{d}t_1\ldots\mathrm{d}t_n))\mathrm{d}r$$

$$= \int_s^t (f_{s\tau}(\emptyset) + \sum_{n=1}^\infty (\int\cdots\int_{s<t_1<\cdots<t_n<\tau} f_{s\tau}(\{t_1,\ldots,t_n\})\mathrm{d}t_1\ldots\mathrm{d}t_n)))\mathrm{d}r$$

$$= \sum_{n=1}^\infty (\int\cdots\int_{s<t_1<\cdots<t_n<t} f_{st_n}(\{t_1,\ldots,t_{n-1}\})\mathrm{d}t_1\ldots\mathrm{d}t_n)$$

which is the right hand side.◇

We have the following remarkable consequence.

2.4.3. THEOREM. *Let* (k_{st}) *be in* $\mathcal{A}_D$. *Then for* $E \in \mathcal{P}_0(D)$, $\alpha \in \mathbb{R}_+$, *and* $\nu = 1, 2, 3$

$$\begin{aligned}
&\int_{S^3} \|(\int_s^t k_{sr} \mathrm{d}m_{\nu,r})(\sigma,\tau,\varrho)\lceil E^{\otimes(\tau\cup\varrho)}\|^\alpha \mathrm{d}\sigma\mathrm{d}\tau\mathrm{d}\varrho \\
&= (c_\nu)^\alpha \int_s^t (\int_{S^3} \|k_{sr}(\sigma,\tau,\varrho)\lceil E^{\otimes(\tau\cup\varrho)}\|^\alpha \mathrm{d}\sigma\mathrm{d}\tau\mathrm{d}\varrho)\mathrm{d}r
\end{aligned} \tag{2.4.2}$$

where

$$c_\nu = \begin{cases} \|\xi\| & \textit{if } \nu = 1, 3 \\ \|T\lceil E\| & \textit{if } \nu = 2. \end{cases}$$

PROOF: We apply Lemma 2.4.2 to

$$f_{st}(\sigma) = \int_{S^2} \|k_{st}(\sigma,\tau,\varrho)a^*(t)(\xi)\lceil E^{\otimes(\tau\cup\varrho)}\|^\alpha \mathrm{d}\tau\mathrm{d}\varrho$$

to obtain

$$\begin{aligned}
&\int_{S^3} \|(\int_s^t k_{sr} \mathrm{d}m_{\nu,r}(\sigma,\tau,\varrho)\lceil E^{\otimes(\tau\cup\varrho)}\|^\alpha \mathrm{d}\sigma\mathrm{d}\tau\mathrm{d}\varrho \\
&= \int_{S\cap[s,t)\setminus S_0} (\int_{S^2} \|k_{s\hat\sigma}(\overline{\sigma},\tau,\varrho)a^*(\hat\sigma)(\xi)\lceil E^{\otimes(\tau\cup\varrho)}\|^\alpha \mathrm{d}\tau\mathrm{d}\varrho\mathrm{d}\sigma\mathrm{d}r \\
&= \|\xi\|^\alpha \int_s^t (\int_{S^3} \|k_{sr}(\sigma,\tau,\varrho)\lceil E^{\otimes(\tau\cup\varrho)}\|^\alpha \mathrm{d}\sigma\mathrm{d}\tau\mathrm{d}\varrho)\mathrm{d}r.
\end{aligned}$$

The other cases are treated in the same manner.◇

Theorem 2.4.4 implies the following.

2.4.3. THEOREM. *If* $(k_{st}) \in \mathcal{A}_D$ *is such that* $\underline{k}_{st} = 0$ *for all* $(s,t) \in T$ *then the operators associated with the kernels* $\int_s^t k_{sr} \mathrm{d}m_{\nu,r}$ *are equal to* 0 *for all* $(s,t) \in T$ *for* $\nu = 1, 2, 3, 4$.

PROOF: By Theorem 2.2.2

$$k_{st}(\sigma,\tau,\varrho)\lceil E^{\otimes(\tau\cup\varrho)} = 0 \tag{2.4.3}$$

for almost all (σ,τ,ϱ). Clearly, this means that

$$\int_s^t k_{sr}(\sigma,\tau,\varrho)\lceil E^{\otimes(\tau\cup\varrho)}\mathrm{d}r = 0$$

for almost all (σ,τ,ϱ). Moreover, (2.4.3) means that for $\nu = 1, 2, 3$

$$\int_s^t (\int_{S^3} \|k_{sr}(\sigma,\tau,\varrho)\lceil E^{\otimes(\tau\cup\varrho)}\|\mathrm{d}\sigma\mathrm{d}\tau\mathrm{d}\varrho)\mathrm{d}r = 0$$

which by Theorem 2.4.2 gives that

$$\int_{S^3} \|(\int_s^t k_{sr} \mathrm{d}m_{\nu,r})(\sigma,\tau,\varrho)\lceil E^{\otimes(\tau\cup\varrho)}\|\mathrm{d}\sigma\mathrm{d}\tau\mathrm{d}\varrho = 0$$

which in turn yields

$$\left(\int_s^t k_{sr} \, dm_{\nu,r}\right)(\sigma, \tau, \varrho) \lceil E^{\otimes(\tau \cup \varrho)} = 0$$

for almost all (σ, τ, ϱ). The theorem now follows from another application of Theorem 2.2.2.$\diamond$

Due to the preceding theorem, we are now in a position to introduce the quantum stochastic integrals

$$\int_s^t \underline{k}_{sr} \, dM_{\nu,r} \tag{2.4.4}$$

for families $(\underline{k}_{st})$ of operators on $\mathcal{E}_D$ in the class $\underline{A}_D$ which is the image of $\mathcal{A}_D$ under the transformation $k_{st} \mapsto \underline{k}_{st}$. We define the integral (2.4.4) to be the image of $\int_s^t k_{sr} \, dm_{\nu,r}$ under this transformation.

According to our introductory remarks our integrals are of Ito type and we may ask if there is something like the Ito product formula in classical stochastic calculus. We wish to express $x_{st}^{(1)} x_{st}^{(2)}$ in terms of quantum stochastic integrals if $x_{st}^{(1)}$ and $x_{st}^{(2)}$ are such integrals.

2.4.5. THEOREM (QUANTUM ITO'S FORMULA). *Let* $(k_{st}^{(1)})$ *and* $(k_{st}^{(2)})$ *be in* $\mathcal{A}_D$ *and let* $\nu_1, \nu_2 \in \{1, 2, 3, 4\}$. *We put*

$$x_{st}^{(i)} = \int_s^t k_{sr}^{(i)} \, dm_{\nu_i, r}$$

for $i = 1, 2$. *Then*

$$x_{st}^{(1)} x_{st}^{(2)} = \int_s^t (k_{sr}^{(1)} x_{sr}^{(2)}) \, dm_{\nu_1, r} + \int_s^t (x_{sr}^{(1)} k_{sr}^{(2)}) \, dm_{\nu_2, r}$$

$$+ \int_s^t (k_{sr}^{(1)} k_{sr}^{(2)})(dm_{\nu_1, r} \, dm_{\nu_2, r}) \tag{2.4.5}$$

where $(dm_{\nu_1, r} \, dm_{\nu_2, r})$ *is to be computed according to the following multiplication table*

	$da_r^*(\xi_2)$	$d\lambda_r(T_2)$	$da_r(\xi_2)$	dr
$da_r^*(\xi_1)$	0	0	0	0
$d\lambda_r(T_1)$	$da_r^*(T_1\xi_2)$	$d\lambda_r(T_1 T_2)$	0	0
$da_r(\xi_1)$	$\langle \xi_1, \xi_2 \rangle dr$	$da_r(T_2^*\xi_1)$	0	0
dr	0	0	0	0

PROOF: We restrict ourselves to the cases $\nu_1 = 3$, $\nu_2 = 1$ and $\nu_1 = 1$, $\nu_2 = 3$, the other cases can be treated similarly. $\nu_1 = 3, \nu_2 = 1$: We have

$$
\begin{aligned}
&(x^{(1)}_{st} x^{(2)}_{st})(\sigma,\tau,\varrho) \\
&= \sum_{\substack{\sigma_1\dot\cup\sigma_2\dot\cup\sigma_3=\sigma\\ \tau_1\dot\cup\tau_2\dot\cup\tau_3=\tau\\ \varrho_1\dot\cup\varrho_2\dot\cup\varrho_3=\varrho}} \int_{S\setminus S_0} k^{(1)}_{s,(\varrho_1\cup\varrho_2\cup\varepsilon)^\wedge}(\sigma_1,\sigma_2\cup\tau_1\cup\tau_2,(\varrho_1\cup\varrho_2\cup\varepsilon)^-)\times \\
&\qquad \times a((\varrho_1\cup\varrho_2\cup\varepsilon)^\wedge)(\xi_1)\times \\
&\qquad\qquad \times k^{(2)}_{s,(\varepsilon\cup\sigma_2\cup\sigma_3)^\wedge}((\varepsilon\cup\sigma_2\cup\sigma_3)^-,\tau_2\cup\tau_3\cup\varrho_2,\varrho_3)\times \\
&\qquad\qquad\qquad \times a^*((\varepsilon\cup\sigma_2\cup\sigma_3)^\wedge)(\xi_2)d\varepsilon \\
&+ \sum_{\substack{\sigma_1\dot\cup\sigma_2\dot\cup\sigma_3=\sigma\\ \tau_1\dot\cup\tau_2\dot\cup\tau_3=\tau\\ \varrho_1\dot\cup\varrho_2\dot\cup\varrho_3=\varrho\\ \sigma_2\cup\sigma_3\neq\emptyset;\varrho_1\cup\varrho_2\neq\emptyset}} k^{(1)}_{s,(\varrho_1\cup\varrho_2)^\wedge}(\sigma_1,\sigma_2\cup\tau_1\cup\tau_2,(\varrho_1\cup\varrho_2)^-)a((\varrho_1\cup\varrho_2)^\wedge)(\xi_1)\times \\
&\qquad \times k^{(2)}_{s,(\sigma_2\cup\sigma_3)^\wedge}((\sigma_2\cup\sigma_3)^-,\tau_2\cup\tau_3\cup\varrho_2,\varrho_3)\times \\
&\qquad\qquad \times a^*((\sigma_2\cup\sigma_3)^\wedge)(\xi_2).
\end{aligned}
\tag{2.4.6}
$$

The maximum $t_0 = (\sigma\dot\cup\tau\dot\cup\varrho)^\wedge$ is either in σ or in τ or in ϱ. It follows from the adaptedness of $(k^{(i)}_{st})$, $i = 1, 2$, that, for $t_0 \in \tau$, the second summand vanishes and the first is equal to

$$
\begin{aligned}
&\sum_{\substack{\sigma_1\dot\cup\sigma_2\dot\cup\sigma_3=\sigma\\ \tau_1\dot\cup\tau_2\dot\cup\tau_3=\tau\\ \varrho_1\dot\cup\varrho_2\dot\cup\varrho_3=\varrho}} \int_{\hat\varepsilon>(\varrho_1\cup\varrho_2\cup\sigma_2\cup\sigma_3)^\wedge} k^{(1)}_{s,\hat\varepsilon}(\sigma_1,\sigma_2\cup\tau_1\cup\tau_2,\varrho_1\cup\varrho_2\cup\bar\varepsilon)a(\bar\varepsilon)(\xi_1)\times \\
&\qquad\qquad \times k^{(2)}_{s,\hat\varepsilon}(\hat\varepsilon\cup\sigma_2\cup\sigma_3,\tau_2\cup\tau_3\cup\varrho_2,\varrho_3)a^*(\hat\varepsilon)(\xi_2)d\varepsilon \\
&= \sum_{\substack{\sigma_1\dot\cup\sigma_2\dot\cup\sigma_3=\sigma\\ \tau_1\dot\cup\tau_2\dot\cup\tau_3=\tau\\ \varrho_1\dot\cup\varrho_2\dot\cup\varrho_3=\varrho}} \int_s^t \Big(\int_S k^{(1)}_{sr}(\sigma_1,\sigma_2\cup\tau_1\cup\tau_2,\varrho_1\cup\varrho_2\cup\varepsilon)\times \\
&\qquad\qquad \times k^{(2)}_{sr}(\varepsilon\cup\sigma_2\cup\sigma_3,\tau_2\cup\tau_3\cup\varrho_2,\varrho_3)d\varepsilon\Big)\langle\xi_1,\xi_2\rangle dr
\end{aligned}
\tag{2.4.7}
$$

which is the second term of the right hand side of (2.4.5) evaluated at (σ,τ,ϱ). Now let $t_0 \in \sigma$. Each of the sums of the right hand side of (2.4.6) splits into the two sums when $t_0 \in \sigma_1 \cup \sigma_2$ and $t_0 \in \sigma_3$. The part for $t_0 \in \sigma_1 \cup \sigma_2$ of the second sum vanishes and the one of the first sum equals the part of (2.4.7) for $t_0 \in \sigma_1 \cup \sigma_2$. The part of (2.4.6) for

$t_0 \in \sigma_3$ gives the part of (2.4.7) for $t_0 \in \sigma_3$ plus the term

$$\sum_{\substack{\sigma_1\cup\sigma_2\cup\sigma_3=\sigma \\ \tau_1\cup\tau_2\cup\tau_3=\tau \\ \varrho_1\cup\varrho_2\cup\varrho_3=\varrho}} \sum_S x^{(1)}_{s,\hat{\sigma}}(\sigma_1, \sigma_2 \cup \tau_1 \cup \tau_2, \varrho_1 \cup \varrho_2 \cup \varepsilon) \times$$

$$\times\, k^{(2)}_{s,\hat{\sigma}}(\varepsilon \cup \sigma_2 \cup \overline{\sigma}_3, \tau_2 \cup \tau_3 \cup \varrho_2, \varrho_3) \mathrm{d}\varepsilon) a^*(\hat{\sigma})(\xi_2)$$

$$= (x^{(1)}_{s,\hat{\sigma}} k^{(2)}_{s,\hat{\sigma}})(\overline{\sigma}, \tau, \varrho) a^*(\hat{\sigma})(\xi_1)$$

$$= \int_s^t (x^{(1)}_{s\tau} k^{(2)}_{s\tau}) \mathrm{d}a^*_\tau(\xi_2)(\sigma, \tau, \varrho)$$

where we made use of the adaptedness of (x_{st}). The case $t_0 \in \varrho$ is symmetric to the case $t_0 \in \sigma$. $\nu_1 = 1$, $\nu_2 = 3$: We have

$$(x^{(1)}_{st} x^{(2)}_{st})(\sigma, \tau, \varrho)$$

$$= \sum_{\substack{\sigma_1\dot\cup\sigma_2\dot\cup\sigma_3=\sigma \\ \tau_1\dot\cup\tau_2\dot\cup\tau_3=\tau \\ \varrho_1\dot\cup\varrho_2\dot\cup\varrho_3=\varrho \\ \sigma_1\neq\emptyset;\sigma_3\neq\emptyset}} \int_S k^{(1)}_{s,\hat{\sigma}_1}(\overline{\sigma}_1, \sigma_2 \cup \tau_1 \cup \tau_2, \varrho_1 \cup \varrho_2 \cup \varepsilon) a^*(\hat{\sigma}_1)(\xi_1) \times$$

$$\times\, k^{(2)}_{s,\hat{\varrho}_3}(\varepsilon \cup \sigma_2 \cup \sigma_3, \tau_2 \cup \tau_3 \cup \varrho_2, \overline{\varrho}_3) a(\hat{\varrho}_3)(\xi_2) \mathrm{d}\varepsilon.$$

The last integral vanishes unless $(\sigma \cup \tau \cup \varrho)^\wedge = \hat{\sigma}_1$ or $(\sigma \cup \tau \cup \varrho)^\wedge = \hat{\varrho}_3$. If we split the sum into the two parts corresponding to these cases we obtain the terms

$$\int_s^t (k^{(1)}_{s\tau} x^{(2)}_{s\tau}) \mathrm{d}a^*(\xi_1)$$

and

$$\int_s^t (x^{(1)}_{s\tau} k^{(2)}_{s\tau}) \mathrm{d}a(\xi_2). \diamond$$

Frequently, we use the 'differential notation'

$$\mathrm{d}x_{st} = k_{st} \mathrm{d}m_{\nu,t},$$

$$x_{ss} = k_0$$

to indicate the fact that

$$x_{st} = k_0 + \int_s^t k_{s\tau} \mathrm{d}m_{\nu,\tau}.$$

For operators Ito's formula reads, in short hand notation,

$$\mathrm{d}(\underline{x}^{(1)}_{st} \underline{x}^{(2)}_{st}) = \underline{x}^{(1)}_{st}(\mathrm{d}\underline{x}^{(2)}_{st}) + (\mathrm{d}\underline{x}^{(1)}_{st})\underline{x}^{(2)}_{st} + (\mathrm{d}\underline{x}^{(1)}_{st})(\mathrm{d}\underline{x}^{(2)}_{st})$$

where all the processes commute with all differentials and the Ito term is computed according to the above table.

2.5. Coalgebra stochastic integral equations. Let $\mathcal{C}$ be a coalgebra and assume that (k_{st}) is given by η, ρ, ϑ and ψ like in Section 2.3. Denote by $\Omega^{(3)}$ the kernel which is concentrated at $(\emptyset, \emptyset, \emptyset)$ with $\Omega^{(3)}(\emptyset, \emptyset, \emptyset) = \mathrm{id}$, i.e. $\Omega^{(3)}$ is the kernel of the identity operator. We have the following theorem.

2.5.1. THEOREM. *The kernels $k_{st}(b)$ form the unique family of kernel-processes $(k_{st}(b))_{(s,t)\in T}$ in $\mathcal{A}_D$ satisfying the equations*

$$\begin{aligned} k_{st}(b) = \epsilon(b)\Omega^{(3)} + \sum_i \int_s^t k_{sr}(b_{(1)})(\mathrm{d}a_r^*(\eta(b_{(2)})) + \mathrm{d}\lambda_r(\rho(b_{(2)}) - \epsilon(b_{(2)})\mathrm{id}) \\ + \mathrm{d}a_r(\vartheta(b_{(2)})) + \psi(b_{(2)})\mathrm{d}r). \end{aligned} \tag{2.5.1}$$

PROOF: Let $b \in \mathcal{C}$ and assume that $(k_{st}(b)) \in \mathcal{A}_D$ satisfies (2.5.1). Then

$$k_{st}(b)(\emptyset, \emptyset, \emptyset) = \epsilon(b)\mathrm{id} + \int_s^r (k_{sr}(\emptyset, \emptyset, \emptyset) \star \psi)(b)\mathrm{d}r.$$

However, the unique solution of this equation is

$$k_{st}(b)(\emptyset, \emptyset, \emptyset) = \varphi_{t-s}(b).$$

Now let $E \in \mathcal{P}_0(D)$ and assume that $k_{st}(b)(\sigma', \tau', \varrho')\lceil E^{\otimes(\tau' \cup \varrho')}$ is determined for $\#(\sigma' \cup \tau' \cup \varrho') \leq n$, $n \in \mathbb{N}$. For disjoint σ, τ, ϱ with $\#(\sigma \cup \tau \cup \varrho) = n + 1$ there are the three cases when $t_0 = (\sigma \cup \tau \cup \varrho)^\wedge$ is an element of σ, τ or ϱ. If $t_0 \in \sigma$, $t_0 < t$, equation (2.5.1) gives

$$\begin{aligned} k_{st}(b)(\sigma, \tau, \varrho)\lceil E^{\otimes(\tau\cup\varrho)} = \sum_i k_{st_0}(b_{(1)})(\overline{\sigma}, \tau, \varrho)a^*(t_0)(\eta(b_{(2)}))\lceil E^{\otimes(\tau\cup\varrho)} \\ + \int_{t_0}^t (k_{sr}(\sigma, \tau, \varrho) \star \psi)(b)\lceil E^{\otimes(\tau\cup\varrho)}\mathrm{d}r. \end{aligned} \tag{2.5.2}$$

By the induction hypothesis, the first part of the left hand side of (2.5.2) is already determined. Moreover, it remains constant for $t > t_0$. It follows that (2.5.2) has a unique solution $k_{st}(b)(\sigma, \tau, \varrho)$ for $t > t_0$. Similarly, if $t_0 \in \tau$ or $t_0 \in \varrho$.

We show that $k_{st}(b)$ given by η, ρ, ϑ and ψ solves (2.5.1). We have for $\sigma \dot{\cup} \tau \dot{\cup} \varrho = \{t_1 < \cdots < t_n\} \subset [s, t)$ and $t_n \in \sigma$

$$\begin{aligned} &\sum_i \int_s^t k_{sr}(b_{(1)})\mathrm{d}a_r^*(\eta(b_{(2)}))(\sigma, \tau, \varrho) \\ &= \sum_i k_{st_n}(b_{(1)})(\overline{\sigma}, \tau, \varrho)a^*(t_n)(\eta(b_{(2)})) \\ &= (\varphi_{t_1-s} \star \kappa(t_1; \sigma, \tau, \varrho) \star \ldots \\ &\qquad \ldots \star \varphi_{t_{n-2}-t_{n-1}} \star \kappa(t_{n-1}; \sigma, \tau, \varrho) \star \varphi_{t_n - t_{n-1}} \star \kappa(t_n; \sigma, \tau, \varrho))(b) \end{aligned}$$

and

$$\int_s^t (k_{sr}(\sigma,\tau,\varrho) \star \psi)(b)\mathrm{d}r$$

$$= \int_{t_n}^t (\varphi_{t_1-s} \star \kappa(t_1;\sigma,\tau,\varrho) \star \ldots \star \varphi_{t_n-t_{n-1}} \star \kappa(t_n;\sigma,\tau,\varrho) \star \varphi_{r-t_n} \star \psi)(b)\mathrm{d}r$$

$$= [(\varphi_{t_1-s} \star \kappa(t_1;\sigma,\tau,\varrho) \star \ldots \star \varphi_{t_n-t_{n-1}} \star \kappa(t_n;\sigma,\tau,\varrho) \star \varphi_{r-t_n})(b)]_{t_n}^t.$$

Similar computations hold for $t_n \in \tau$ and $t_n \in \varrho$. Since $k_{st}(b)(\emptyset,\emptyset,\emptyset) = \varphi_{t-s}(b)$, it follows that $k_{st}(b)$ satisfies (2.5.1).◇

The following stronger version of the uniqueness statement in Theorem 2.5.1 will be needed.

2.5.2. PROPOSITION. *Let $(\tilde{k}_{st}(b))_{(s,t)\in T}$ be in $\mathcal{A}_D$. Suppose that for each $(s,t) \in T$ there is a subset N_{st} of $\mathcal{S}^3$ of measure 0 such that $\tilde{k}_{st}(b)$ satisfies (2.5.1) for all $(\sigma,\tau,\varrho) \notin N_{st}$. Then there is a subset N of $\mathcal{S}^3$ of measure 0 such that*

$$\tilde{k}_{st}(b)(\sigma,\tau,\varrho) = k_{st}(b)(\sigma,\tau,\varrho)$$

for all $(\sigma,\tau,\varrho) \notin N$ and for all $(s,t) \in T$.

PROOF: Suppose that

$$\begin{aligned} &\tilde{k}_{st}(b)(\sigma,\tau,\varrho) \\ &= \epsilon(b)\Omega^{(3)}(\sigma,\tau,\varrho) + \sum_i \int_s^t \tilde{k}_{sr}(b_{1i}) \times \\ &\qquad \times (\mathrm{d}a_r^*(\eta(b_{2i})) + \mathrm{d}\lambda_r(\rho(b_{2i}) - \epsilon(b_{2i})\mathrm{id}) \\ &\qquad + \mathrm{d}a_r(\vartheta(b_{2i})) + \psi(b_{2i})\mathrm{d}r)(\sigma,\tau,\varrho) \end{aligned} \tag{2.5.3}$$

for all $(\sigma,\tau,\varrho) \notin N_{st}$. Call the right hand side of this equation $(I\tilde{k})_{st}(b)(\sigma,\tau,\varrho)$. Then $(I\tilde{k})_{st}(b)$ is in $\mathcal{A}_D$ and it agrees with $\tilde{k}_{st}(b)$ almost surely for each fixed $(s,t) \in T$. It follows that we can replace $\tilde{k}_{st}(b)$ by $(I\tilde{k})_{st}(b)$, i.e. we can assume that $\tilde{k}_{st}(b)(\sigma,\tau,\varrho)$ is continuous in s and in t for each fixed (σ,τ,ϱ) but for the point $(\sigma \cup \tau \cup \varrho)^\wedge$. By a standard argument, it follows that there is a set $N_0 \subset \mathcal{S}^3$ of measure 0 such that (2.5.3) holds for all $(\sigma,\tau,\varrho) \notin N_0$ and all $(s,t) \in T$. Define the sets N_n inductively by

$$N_{n+1} = \{(\sigma,\tau,\varrho) \in \mathcal{S}^3 : \sigma,\tau,\varrho \text{ disjoint and } (\sigma,\tau,\varrho)^- \in N_n\}$$

where $(\sigma,\tau,\varrho)^-$ means $(\overline{\sigma},\tau,\varrho)$, $(\sigma,\overline{\tau},\varrho)$ or $(\sigma,\tau,\overline{\varrho})$ depending on whether the maximum of $\sigma \cup \tau \cup \varrho$ is in σ, τ or ϱ. Then the N_n have measure 0. Using the same induction argument as in the beginning of the proof of Theorem 2.5.1, the last statement of the theorem follows if we put $N = \bigcup_{n\in\mathbb{N}} N_n$.◇

We come to the

PROOF OF THEOREM 2.3.3: We define the linear mapping $L_{st} : \mathcal{B} \otimes \mathcal{B} \to \mathcal{A}_D$ by

$$L_{st}(b \otimes c) = k_{st}(b)k_{st}(c).$$

Using Ito's formula, we obtain

$$\begin{aligned}
&L_{st}(b\otimes c)\\
&=\epsilon(bc)\Omega^{(3)}+\sum_i\int_s^t(k_{sr}(b_{1i})k_{sr}(c))\times\\
&\qquad\times(\mathrm{d}a_r^*(\eta(b_{2i}))+\mathrm{d}\lambda_r(\rho)b_{2i})-\epsilon(b_{2i})\mathrm{id})\\
&\qquad+\mathrm{d}a_r(\vartheta(b_{2i}))+\psi(b_{2i})\mathrm{d}r)\\
&\quad+\sum_j\int_s^t(k_{sr}(b)k_{sr}(c_{1j}))\times\\
&\qquad\times(\mathrm{d}a_r^*(\eta(c_{2j}))+\mathrm{d}\lambda_r(\rho(c_{2j})-\epsilon(c_{2j})\mathrm{id})\\
&\qquad+\mathrm{d}a_r(\vartheta(c_{2j})+\psi(c_{2j})\mathrm{d}r)\\
&\quad+\sum_{i,j}\int_s^t(k_{sr}(b_{1i})k_{sr}(c_{1j}))\times\\
&\qquad\times(\mathrm{d}a_r^*(\rho(b_{2i})\eta(c_{2j})-\epsilon(b_{2i})\eta(c_{2j}))\\
&\qquad+\mathrm{d}\lambda_r((\rho(b_{2i})-\epsilon(b_{2i}))(\rho(c_{2j})-\epsilon(c_{2j})))\\
&\qquad+\mathrm{d}a_r(\rho(c_{2j})^*\vartheta(b_{2i})-\epsilon(c_{2j})\vartheta(b_{2i}))+\langle\vartheta(b_{2i}),\eta(c_{2j})\rangle\mathrm{d}r).
\end{aligned}$$

Now we use the additional assumptions on η, ρ, ϑ and ψ and have

$$\begin{aligned}
&L_{st}(b\otimes c)\\
&=\epsilon(bc)\Omega^{(3)}+\sum_{i,j}\int_s^t L_{sr}(b_{1i}\otimes c_{1j})\times\\
&\qquad\times(\mathrm{d}a_r^*(\eta(b_{2i}c_{2j}))+\mathrm{d}\lambda_r(\rho(b_{2i}c_{2j})-\epsilon(b_{2i}c_{2j})\mathrm{id})\\
&\qquad+\mathrm{d}a_r(\vartheta(b_{2i}c_{2j}))+\psi(b_{2i}c_{2j})\mathrm{d}r).
\end{aligned}$$

This means that $L_{st}=k_{st}^{(\eta',\rho',\vartheta',\psi')}$ with $\eta'(b\otimes c)=\eta(bc)$, $\rho'(b\otimes c)=\rho(bc)$, $\vartheta'(b\otimes c)=\vartheta(bc)$ and $\psi'(b\otimes c)=\psi(bc)$. On the other hand,

$$(b\otimes c)\mapsto k_{st}(bc)$$

also is equal to $k_{st}^{(\eta',\rho',\vartheta',\psi')}$ and

$$k_{st}(b)k_{st}(c)=k_{st}(bc)$$

by Theorem 2.5.1.◇

We summarize the equations (2.5.1) to

$$k_{st}=\epsilon\Omega^{(3)}+\int_s^t k_{sr}\star(\mathrm{d}a_r^*\circ\eta+\mathrm{d}\lambda_r\circ(\rho-\epsilon)+\mathrm{d}a_r\circ\vartheta+\psi\mathrm{d}r)$$

or

$$\underline{k}_{st}=\epsilon\mathrm{id}+\int_s^t\underline{k}_{sr}\star(\mathrm{d}A_r^*\circ\eta+\mathrm{d}\Lambda_r\circ(\rho-\epsilon)+\mathrm{d}A_r\circ\vartheta+\psi\mathrm{d}r)$$

and the representation theorem now reads as follows.

2.5.3. THEOREM. *Let $\mathcal{B}$ be a $*$-bialgebra and let η, ρ and ψ satisfy the conditions of Theorem 2.3.5. Then the quantum stochastic differential equations*

$$\mathrm{d}j_{st} = j_{st} \star (\mathrm{d}A_t^* \circ \eta + \mathrm{d}\Lambda_t \circ (\rho - \epsilon) + \mathrm{d}A_t \circ \tilde{\eta} + \psi \mathrm{d}t) \tag{2.5.4}$$

with the initial conditions

$$j_{ss} = \epsilon\,\mathrm{id}$$

have unique solutions $(j_{st}(b))$ in $\underline{A}_D$. Moreover, in the vacuum state, the j_{st} form a Bose white noise on $\mathcal{B}$ with generator ψ. Conversely, a Bose white noise on $\mathcal{B}$ with generator ψ is equivalent to $(j_{st})_{(s,t)\in T}$ with $\eta = \eta_\psi$ and $\rho = \rho_\psi$.$\diamondsuit$

PROOF: Assume that $(\tilde{j}_{st}) \in \underline{A}_D$ satisfies (2.5.4). Let $(\tilde{k}_{st})$ be an element in $\mathcal{A}_D$ such that $\underline{\tilde{k}}_{st} = \tilde{j}_{st}$ for all $(s,t) \in T$. Then $(\tilde{k}_{st})$ must satisfy (2.5.1) almost everywhere for each $(s,t) \in T$. By Prop. 2.5.2. this implies that $\tilde{k}_{st} = k_{st}$ a.e. and thus $\tilde{j}_{st} = j_{st}$.$\diamondsuit$

Whereas the quantum Ito integrals are *forward integrals* we can define the *backward integrals* by

$$\left(\int_s^t \mathrm{d}a_\tau^*(\xi) k_{\tau t}\right)(\sigma, \tau, \varrho) = \begin{cases} k_{\check{\sigma}t}(\underline{\sigma}, \tau, \varrho) a^*(\check{\sigma})(\xi) & \text{if } \sigma \neq \emptyset \text{ and } t \geq \check{\sigma} > s \\ 0 & \text{otherwise,} \end{cases}$$

and similarly for the other types of integrals. With the obvious notation we obtain that our solutions $j_{st}(b)$ of the equations (2.5.4) also satisfy the backward equations

$$\mathrm{d}j_{st} = (\mathrm{d}A_s^* \circ \eta + \mathrm{d}\Lambda_s \circ (\rho - \epsilon) + \mathrm{d}A_s \circ \tilde{\eta} + \psi \mathrm{d}s) \star j_{st} \tag{2.5.5}$$

with the initial condition $j_{tt} = \epsilon\,\mathrm{id}$.

3. Symmetrization.

We use a semi-direct product construction to 'reduce' (γ, α)-bialgebras to symmetric bialgebras, i.e. (γ, α)-bialgebras with trivial double-module structure. This leads to a description of (γ, α)-white noise in terms of symmetric white noise.

3.1. Symmetrization of bialgebras. Let $\mathbb{L}$ be a $*$-group. The free vector space $\mathbb{C}\mathbb{L}$ spanned by $\mathbb{L}$ is a symmetric, cocommutative $*$-Hopf algebra; see Example 1 of Section 1.8. The $*$-algebra $\mathbb{C}\mathbb{L}$ is $\mathbb{L}$-graded if we put $(\mathbb{C}\mathbb{L})^{(\lambda)}$ equal to the 1-dimensional subspace of $\mathbb{C}\mathbb{L}$ spanned by λ. For a vector space V we denote by $V_{\mathbb{L}}$ the vector space $\mathbb{C}\mathbb{L} \otimes V$. There are two natural linear mappings

$$r_{\mathbb{L}} : V_{\mathbb{L}} \to V, \quad l_{\mathbb{L}} : V \to V_{\mathbb{L}}$$

given by

$$r_{\mathbb{L}}(\lambda \otimes v) = v, \quad l_{\mathbb{L}}(v) = \mathrm{e} \otimes v$$

respectively. We have $r_{\mathbb{L}} \circ l_{\mathbb{L}} = \mathrm{id}$. If V is a module $*$-vector space, $l_{\mathbb{L}}$ is an embedding of the $*$-vector space V into the $*$-vector space $V_\alpha = \mathbb{C}\mathbb{L} \otimes_{(\gamma,\alpha)} V$. If $\mathcal{A}$ is a module algebra, $l_{\mathbb{L}}$ is an embedding of the algebra $\mathcal{A}$ into the semi-direct product algebra $\mathcal{A}_\alpha = \mathbb{C}\mathbb{L} \otimes_{(\gamma,\alpha)} \mathcal{A}$. In this case, we also have the embedding of $\mathbb{C}\mathbb{L}$ into $\mathcal{A}_\alpha$ given by $\lambda \mapsto \lambda \otimes \mathbf{1}$. We identify the algebras $\mathbb{C}\mathbb{L}$ and $\mathcal{A}$ with there images in $\mathcal{A}_\alpha$. Then $\mathcal{A}_\alpha$ is the free product $\mathbb{C}\mathbb{L} \star \mathcal{A}$ of algebras divided by the ideal generated by the elements $\lambda a - (\lambda.a)\lambda$, $\lambda \in \mathbb{L}$, $a \in \mathcal{A}$. If $\mathcal{A}$ is a module $*$-algebra, this ideal is a $*$-ideal. If we divide the free product $\mathbb{C}\mathbb{L} \star \mathcal{A}$ of $*$-algebras by the ideal we obtain the *$*$-algebra* $\mathcal{A}_\alpha = \mathbb{C}\mathbb{L} \otimes_{(\gamma,\alpha)} \mathcal{A}$. Next we observe that $V_{\mathbb{L}}$ is a module vector space with respect to

$$\lambda'(\lambda \otimes v) = (\lambda'\lambda) \otimes v.$$

For a graded vector space V and a vector space W we introduce the linear mapping

$$\iota_\gamma : V \otimes W \to V \otimes W_{\mathbb{L}}$$

which on $V^{(\lambda)} \otimes W$ is the tensor product of the identity with the multiplication by λ, i.e.

$$\iota_\gamma(v \otimes w) = v \otimes \mathrm{d}(v)(l_{\mathbb{L}} w) = v \otimes \mathrm{d}(v) \otimes w.$$

We have the following proposition.

3.1.1. PROPOSITION.

(i) *Let V be a comodule and let W be a module $*$-vector space. The mapping ι_γ is a $*$-mapping from $V \otimes_{(\gamma,\alpha)} W$ to the ordinary tensor product $V \otimes W_\alpha$ of involutive vector spaces.*

(ii) *Let $\mathcal{A}$ be a comodule and let $\mathcal{B}$ be a module algebra. The mapping ι_γ is an algebra homomorphism from $\mathcal{A} \otimes_{(\gamma,\alpha)} \mathcal{B}$ to the ordinary tensor product $\mathcal{A} \otimes \mathcal{B}_\alpha$ of algebras.*

PROOF: (i) follows from

$$\begin{aligned}\iota_\gamma((v\otimes w)^*) &= \iota_\gamma(v^*\otimes w^*.\mathrm{d}(v^*))\\ &= v^*\otimes \mathrm{d}(v^*)\otimes w^*.\mathrm{d}(v)^*\\ &= v^*\otimes(\mathrm{d}(v)\otimes w)^*\\ &= (\iota_\gamma(v\otimes w))^*.\end{aligned}$$

(ii): We have

$$\begin{aligned}\iota_\gamma((a\otimes b)(a'\otimes b')) &= aa'\otimes \mathrm{d}(aa')\otimes(b.\mathrm{d}(a'))b'\\ &= aa'\otimes(\mathrm{d}(a)\otimes b)(\mathrm{d}(a')\otimes b')\\ &= \iota_\gamma(a\otimes b)\iota_\gamma(a'\otimes b').\diamond\end{aligned}$$

Now let $\mathcal{C}$ be a graded coalgebra. We introduce the linear mappings

$$\Delta_\gamma : \mathcal{C}_{\mathbb{L}} \to \mathcal{C}_{\mathbb{L}} \otimes \mathcal{C}_{\mathbb{L}}$$

and

$$\epsilon_\gamma : \mathcal{C}_{\mathbb{L}} \to \mathbb{C}$$

by setting

$$\Delta_\gamma(\lambda\otimes c) = (\lambda\otimes\lambda)((l_{\mathbb{L}}\otimes \mathrm{id}\otimes \mathrm{id})\circ\iota_\gamma\circ\Delta)(c)$$

and

$$\epsilon_\gamma(\lambda\otimes c) = \epsilon(c)$$

respectively.

3.1.2. PROPOSITION.

(i) *The triplet $(\mathcal{C}_{\mathbb{L}}, \Delta_\gamma, \epsilon_\gamma)$ forms a coalgebra. We denote it by $\mathcal{C}_\gamma$.*

(ii) *If $\mathcal{B}$ is a (γ,α)-bialgebra, then $(\mathcal{B}_{\mathbb{L}}, \Delta_\gamma, \epsilon_\gamma)$ turns the algebra $\mathcal{B}_\alpha$ into a symmetric bialgebra. We denote it by $\mathcal{B}_{(\gamma,\alpha)}$. If $\mathcal{B}$ is an involutive (γ,α)-bialgebra, $\mathcal{B}_{(\gamma,\alpha)}$ is an involutive symmetric bialgebra, if $\mathcal{B}$ is a (γ,α)-Hopf algebra, $\mathcal{B}_{(\gamma,\alpha)}$ is a symmetric Hopf algebra, and if $\mathcal{B}$ is an involutive (γ,α)-Hopf algebra, $\mathcal{B}_{(\gamma,\alpha)}$ is an involutive symmetric Hopf algebra.*

PROOF: (i): We show that Δ_γ is coassociative. We have for $\lambda\in\mathbb{L}$ and homogeneous $b\in\mathcal{B}$

$$\begin{aligned}(\mathrm{id}\otimes\Delta_\gamma)\circ\Delta_\gamma(\lambda\otimes b) &= \sum \lambda\otimes b_{(1)}\otimes\lambda\mathrm{d}(b_{(1)})\otimes b_{(2)}\otimes\lambda\mathrm{d}(b_{(1)}b_{(2)})\otimes b_{(3)}\\ &= (\Delta_\gamma\otimes\mathrm{id})\circ\Delta_\gamma(\lambda\otimes b).\end{aligned}$$

Next,

$$\begin{aligned}((\epsilon_\gamma\otimes\mathrm{id})\circ\Delta_\gamma)(\lambda\otimes b) &= \sum \epsilon(b_{(1)})\lambda\mathrm{d}(b_{(1)})\otimes b_{(2)}\\ &= \sum \epsilon(b_{(1)})\lambda\otimes b_{(2)}\\ &= \lambda\otimes b,\end{aligned}$$

because ϵ is $\mathbb{L}$-even, and

$$((\mathrm{id} \otimes \epsilon_{\mathbb{L}}) \circ \Delta_{\mathbb{L}})(\lambda \otimes b) = \sum \lambda \otimes b_{(1)}\epsilon(b_{(2)}) = \lambda \otimes b$$

which proves that ϵ_γ is a counit.

(ii): Let $\mathcal{B}$ be a (γ, α)-bialgebra. We must show that Δ_γ and ϵ_γ are algebra homomorphisms. Since $\Delta_\gamma \lceil \mathbb{C}\mathbb{L}$ and $\Delta_\gamma \lceil \mathcal{B} = \iota_\gamma \circ \Delta$ are algebra homomorphism there is a unique algebra homomorphism $\tilde{\Delta}_\gamma$ from $\mathbb{C}\mathbb{L} \star \mathcal{B}$ to $\mathcal{B}_\alpha \otimes \mathcal{B}_\alpha$ such that $\tilde{\Delta}_\gamma \lceil \mathbb{C}\mathbb{L} = \Delta_\gamma$ and $\tilde{\Delta}_\gamma \lceil \mathcal{B} = \Delta_\gamma$. We have

$$\tilde{\Delta}_\gamma(\lambda b) = \sum \lambda b_{(1)} \otimes \lambda \mathrm{d}(b_{(1)}) b_{(2)}$$

and

$$\tilde{\Delta}_\gamma((\lambda.b)\lambda) = \sum (\lambda.b_{(1)})\lambda \otimes (\lambda \mathrm{d}(b_{(1)})\lambda^{-1})(\lambda.b_{(2)})\lambda$$

which shows that $\tilde{\Delta}_\gamma$ gives rise to an algebra homomorphism from $\mathcal{B}_\alpha$ to $\mathcal{B}_\alpha \otimes \mathcal{B}_\alpha$. Clearly, this homomorphism must agree with Δ_γ. Using the fact that ϵ is α-invariant, the same argument yields that ϵ_γ is an algebra homomorphism. Our arguments go through for involutive (γ, α)-bialgebras if we replace 'algebra homomorphism' by '*-algebra homomorphism'. Now suppose that S is an antipode for the (γ, α)-bialgebra $\mathcal{B}$. By Prop. 1.7.3 the linear mapping S_γ from $\mathcal{B}$ to $\mathcal{B}_\alpha$ which sends b to $S(b)\mathrm{d}(b)^{-1}$ is an anti-homomorphism, and there is a unique anti-homomorphism $\tilde{S}_\gamma : \mathbb{C}\mathbb{L} \star \mathcal{B} \to \mathcal{B}_\alpha$ such that $\tilde{S}_\gamma \lceil \mathbb{L}$ is the forming of inverses and $\tilde{S}_\gamma \lceil \mathcal{B} = S_\gamma$. Moreover,

$$\begin{aligned}\tilde{S}_\gamma(\lambda b) &= S(b)\mathrm{d}(b)^{-1}\lambda^{-1} \\ &= S(b)\lambda^{-1}\lambda\mathrm{d}(b)^{-1}\lambda^{-1}\end{aligned}$$

and

$$\tilde{S}_\gamma((\lambda.b)\lambda) = \lambda^{-1}S(b.\lambda^{-1})\lambda\mathrm{d}(b^{-1})\lambda^{-1}$$

which shows that $\tilde{S}_\gamma$ gives rise to a linear algebra anti-homomorphism S_γ on $\mathcal{B}_\alpha$. We show that S_γ is an antipode for $\mathcal{B}_{(\gamma,\alpha)}$. We have

$$\begin{aligned}(M \circ (S_\gamma \otimes \mathrm{id}) \circ \Delta_\gamma)(\lambda b) &= \sum S_\gamma(\lambda b_{(1)})\lambda \mathrm{d}(b_{(1)}) b_{(2)} \\ &= \sum S(b_{(1)}) b_{(2)} \\ &= \epsilon_\gamma(\lambda b)\mathbf{1}\end{aligned}$$

and

$$\begin{aligned}(M \circ (\mathrm{id} \otimes S_\gamma) \circ \Delta_\gamma)(\lambda b) &= \sum \lambda b_{(1)} S_\gamma(\lambda \mathrm{d}(b_{(1)}) b_{(2)}) \\ &= \sum \lambda b_{(1)} S(b_{(2)}) \mathrm{d}(b)^{-1}\lambda^{-1} \\ &= \epsilon(b)\mathbf{1} \\ &= \epsilon_\gamma(\lambda b)\mathbf{1}\end{aligned}$$

where we used that ϵ is α-invariant. Suppose that $\mathcal{B}$ is an involutive (γ,α)-Hopf algebra. Then

$$\begin{aligned}S_\gamma(S_\gamma((\lambda b)^*)^*) &= S_\gamma(((\lambda^*)^{-1}S(b^*)\mathrm{d}(b^*)^{-1})^*)\\ &= S_\gamma(\mathrm{d}(b)^{-1}S(b^*)^*\lambda^{-1})\\ &= \lambda S(S(b^*)^*)\mathrm{d}(b)\mathrm{d}(b)^{-1}\\ &= \lambda b\end{aligned}$$

which proves that $\mathcal{B}_{(\gamma,\alpha)}$ is an involutive Hopf algebra.$\diamondsuit$

The relationship between a (γ,α)-bialgebra $\mathcal{B}$ and its symmetrization $\mathcal{B}_{(\gamma,\alpha)}$ can be expressed by saying that the diagrams

$$\begin{array}{ccc}\mathcal{B} & \xrightarrow{\ \imath_{\mathbb{L}}\ } & \mathcal{B}_{(\gamma,\alpha)}\\ \downarrow{\scriptstyle\Delta} & & \downarrow{\scriptstyle\Delta_\gamma}\\ \mathcal{B}\otimes\mathcal{B} & \xrightarrow{(\imath_{\mathbb{L}}\otimes\mathrm{id}\otimes\mathrm{id})\circ\iota_\gamma} & \mathcal{B}_{(\gamma,\alpha)}\otimes\mathcal{B}_{(\gamma,\alpha)}\end{array}$$

and

$$\begin{array}{ccc}\mathcal{B} & \xleftarrow{\ r_{\mathbb{L}}\ } & \mathcal{B}_{(\gamma,\alpha)}\\ \downarrow{\scriptstyle\Delta} & & \downarrow{\scriptstyle\Delta_\gamma}\\ \mathcal{B}\otimes\mathcal{B} & \xleftarrow{\ r_{\mathbb{L}}\otimes r_{\mathbb{L}}\ } & \mathcal{B}_{(\gamma,\alpha)}\otimes\mathcal{B}_{(\gamma,\alpha)}\end{array}$$

commute.

There seems to be no reason why we should prefer (γ,α) rather than (α,γ). Indeed, one also can consider the tensor product $\mathcal{B}\otimes_{(\alpha,\gamma)}\mathbb{CM}$ and the analogues of Prop. 3.1.1 and Prop. 3.1.2 hold. The choice of (γ,α) is adjusted to left white noise. However, if $\mathcal{B}$ is an involutive (α,γ)-bialgebra we can form the symmetric $*$-bialgebra $\mathcal{B}_{(\alpha,\gamma)} = ((\mathcal{B}^{\mathrm{op}})_{(\gamma,\alpha)})^{\mathrm{op}}$. It is immediate to check that the comultiplication Δ^γ of $\mathcal{B}_{(\alpha,\gamma)}$ is given by

$$\Delta^\gamma(b) = \sum b_{(1)}\mathrm{d}(b_{(2)})\otimes b_{(2)}$$

for $b\in\mathcal{B}$. In the commutation factor case, we write $\mathcal{B}_q$ and $\mathcal{B}^q$ for $\mathcal{B}_{(\gamma,\alpha)}$ and $\mathcal{B}_{(\alpha,\gamma)}$ respectively. At the end of this chapter we will see that, while $\mathcal{B}_{(\gamma,\alpha)}$ is the suitable symmetrization for left white noise, $\mathcal{B}_{(\alpha,\gamma)}$ is the natural symmetrization for right white noise.

We form the symmetrizations for some of the examples of 1.8.

Let V be a double-module $*$-vector space. The comultiplication for the symmetrization $\mathcal{T}(V)_{(\gamma,\alpha)}$ of the involutive (γ,α)-bialgebra $\mathcal{T}(V)$ is given by

$$\Delta_\gamma v = \mathbf{1}\otimes v + v\otimes\mathrm{d}(v)$$

for $v\in V$, i.e. v is $(\mathbf{1},\mathrm{d}(v))$-primitive. Moreover, an antipode of $\mathcal{T}(V)_{(\gamma,\alpha)}$ is given by $S(v) = -v\mathrm{d}(v)^{-1}$.

We turn to our special example when V is spanned by x and x^* and we are in the commutation factor case, $q \in \mathbb{C}_0$, with the $\mathbb{L}$-$\mathbb{M}$-graduation given by $\mathbb{L} = (\mathbb{Z} \times \mathbb{Z})^{\text{twist}}$, $\mathbb{M} = \mathbb{Z}^{\text{inv}}$. If one specializes to a $\mathbb{L}_q$-$\mathbb{M}_q$-graduation one has to add certain relations for y and y^*. The $*$-Hopf algebra $\mathcal{T}(V)_q$ can be described as the $*$-algebra generated by x and indeterminates y, y^{-1} with the relations

$$yx = qxy, \quad y^*x = \bar{q}xy^*$$
$$[y, y^*] = 0, \quad yy^{-1} = y^{-1}y = \mathbf{1},$$

and where y is assumed to be group-like and x $(\mathbf{1}, y)$-primitive. This is exactly the deformation $\mathbf{H}^{(0)}_{1;q}(2)$ of the subgroup $\mathbf{H}^{(0)}(2)$ of the Heisenberg group. Going one step further, we form the associated $*$-Hopf algebra $(\mathbf{H}^{(0)}_{q;\alpha}(2))^q$ of the deformation $\mathbf{H}^{(0)}_{q;\alpha}(2)$ of $\mathbf{H}^{(0)}(2)$ which turns out to be equal to the deformation $\mathbf{H}_{1;q,\alpha}(2)$ of the Heisenberg group $\mathbf{H}(2)$. In fact, this is analogous to the commutative situation. Let us look at the commutative and symmetric tensor $*$-Hopf algebra $\mathbb{C}[x, x^*] = \text{Env}(\mathbb{R}_{\mathbb{C}})$ which we regard as an involutive q-Hopf algebra with $q = 1$ and $\mathbb{L} = (\mathbb{Z} \times \mathbb{Z})^{\text{twist}}$, $\mathbb{M} = \mathbb{Z}^{\text{inv}}$. If we apply our construction we obtain $(\mathbb{C}[x, x^*])_q = \mathcal{H}^{(0)}[2]$, and, applying it once more but in the other direction, to $\mathcal{H}^{(0)}[2]$ we obtain $\mathcal{H}[2]$ so that $\mathcal{H}[2] = ((\mathbb{C}[x, x^*])_q)^q$.

The associated $*$-bialgebra $(\mathcal{M}\langle n; q\rangle)_q$ is the $*$-algebra generated by x_{kl}, $k, l = 1, \ldots, n$, and y, y^{-1} with the relations

$$yx_{kl} = qx_{kl}y, \quad y^*x_{kl} = \bar{q}x_{kl}y^*$$
$$[y, y^*] = 0, \quad yy^{-1} = y^{-1}y = \mathbf{1},$$

and such that y and y^{-1} are group-like and

$$\Delta_\gamma x_{kl} = \sum_{m=1}^{n} x_{km} \otimes y^{k-m} x_{ml}.$$

The next proposition will yield a simple relationship between multiplicative q-white noise and multiplicative symmetric white noise.

3.1.3. PROPOSITION. *The $*$-algebra homomorphism*

$$\Theta : \mathcal{M}\langle n; 1\rangle \to (\mathcal{M}\langle n; q\rangle)_q$$

given by

$$\Theta x_{kl} = y^{-k} x_{kl}$$

is a coalgebra embedding.

Moreover, Θ respects the unitarity relations (1.8.1), so that Θ gives rise to a $*$-bialgebra embedding of $\mathcal{U}\langle n; 1\rangle$ into $(\mathcal{U}\langle n; q\rangle)_q$.

PROOF: We have

$$(\Delta_\gamma \circ \Theta)(x_{kl}) = (y^{-k} \otimes y^{-k})(\sum_{m=1}^{n} x_{km} \otimes y^{k-m} x_{ml})$$
$$= \sum_{m=1}^{n} y^{-k} x_{km} \otimes y^{-m} x_{ml}$$
$$= ((\Theta \otimes \Theta) \circ \Delta)(x_{kl}).$$

Next we have

$$\Theta(\sum_{m=1}^{n} x_{km} x_{lm}^*) = \sum_{m=1}^{n} y^k x_{km} x_{lm}^* y^l$$
$$= y^{-k}(\sum_{m=1}^{n} x_{km} x_{lm}^* - \epsilon_{kl} 1) y^l. \diamond$$

3.2. Schoenberg correspondence. The following result goes back to Schoenberg [69]; cf. also [17, 63]. Let X be a set and L a kernel on X, i.e. L is a complex-valued function on $X \times X$. A kernel L is said to be positive definite if

$$\sum_{i,j} \overline{z_i} z_j L(x_i, x_j) \geq 0$$

for all choices of $n \in \mathbb{N}$ and $x_1, \ldots, x_n \in X$, $z_1, \ldots, z_n \in \mathbb{C}$. Then the (pointwise) exponentials $\exp(tA)$ of a kernel A are positive definite for all $t \geq 0$ if and only if A satisfies the conditions

$$A(x, y) = \overline{A(y, x)} \qquad \text{for all } x, y \in X$$

and

$$\sum_{i,j} \overline{z_i} z_j A(x_i, x_j) \geq 0 \tag{3.2.1}$$

for all choices of $n \in \mathbb{N}$, $x_1, \ldots, x_n \in X$ and $z_1, \ldots, z_n \in \mathbb{C}$ with $\sum_i z_i = 0$. A kernel satisfying condition (3.2.1) is called conditionally positive. The correspondence between 1-parameter semi-groups of positive-definite kernels and conditionally positive-definite kernels is often called *Schoenberg correspondence.* The principle of Schoenberg type correspondence plays an important role in mathematics. In this section, we treat a generalized Schoenberg correspondence where the pointwise exponentials are replaced by convolution exponentials with the convolution coming from a general coalgebra structure.

Given a conditionally positive, hermitian linear functional ψ with $\psi(1) = 0$ on a symmetric $*$-bialgebra $\mathcal{B}$ we constructed a white noise on $\mathcal{B}$ with generator ψ in Chapter 2. Since $\varphi_t = \exp_\star t\psi$ are the 1-dimensional distributions of this white noise, it follows that φ_t are states. This establishes a Schoenberg correspondence for 1-parameter convolution semi-groups of states. We generalize this result in two directions. We consider positive sesquilinear forms on coalgebras which generalize the concept of states on $*$-bialgebras. Then we also allow a non-trivial double-module structure to be present.

Let V be a double-module vector space. We form the conjugate vector space $\overline{V}$ of V which as a set is $\{\overline{v} : v \in V\}$. The addition and the scalar multiplication are defined by

$$\overline{v} + \overline{w} = \overline{v + w}, \quad z\overline{v} = \overline{\overline{z}v}, \quad v, w \in V, z \in \mathbb{C}.$$

A graduation of $\overline{V}$ is given by the family $(\overline{V^{(\lambda^*)}})_{\lambda \in \mathbb{L}}$. The graded vector space $\overline{V}$ becomes a double-module vector space if we put

$$\lambda.\overline{v} = \overline{v.\lambda^*}.$$

Let V and W be two double-module vector spaces. An isomorphism Θ from $\overline{\mathbf{L}(V, W)}$ to $\mathbf{L}(\overline{V}, \overline{W})$ is given by

$$(\Theta\overline{R})(\overline{v}) = \overline{R(v)}, \qquad R \in \mathbf{L}(V, W), v \in V.$$

We identify the elements of $\overline{\mathbf{L}(V, W)}$ and $\mathbf{L}(\overline{V}, \overline{W})$. The conjugate vector space $\overline{V \otimes W}$ of the tensor product is identified with $\overline{V} \otimes \overline{W}$ via the map

$$\overline{v \otimes w} \mapsto \overline{v} \otimes \overline{w}.\mathrm{d}(\overline{v}).$$

The vector space $\overline{\mathbb{C}}$ is identified with $\mathbb{C}$ in the obvious way. One checks that the conjugate $(\overline{C}, \overline{\Delta}, \overline{\epsilon})$ of a double-module coalgebra (C, Δ, ϵ) again is a double-module coalgebra.

3.2.1. PROPOSITION. *Let C be a double-module coalgebra. Then the (γ, α)-tensor product $\overline{C} \otimes_{(\gamma,\alpha)} C$ of coalgebras is turned into an involutive (γ, α)-coalgebra if we define the involution on $\overline{C} \otimes C$ by*

$$(\overline{c} \otimes d)^* = \overline{d} \otimes c \text{ for } c, d \in C.$$

PROOF: Clearly, with this involution, $\overline{C} \otimes C$ becomes a double-module $*$-vector space. Moreover,

$$(\overline{\epsilon} \otimes \epsilon)((\overline{c} \otimes d)^*) = \overline{\epsilon(d)}\epsilon(c) = \overline{\overline{\epsilon(c)}\epsilon(d)}$$

which shows that $\overline{\epsilon} \otimes \epsilon$ is a $*$-map. It remains to show that $\overline{\Delta} \otimes_{(\gamma,\alpha)} \Delta$ is a $*$-map, too. We have

$$\begin{aligned}&(\overline{\Delta} \otimes_{(\gamma,\alpha)} \Delta)((\overline{c} \otimes d)^*)\\ &= \sum \overline{d_{(1)}} \otimes c_{(1)} \otimes \overline{d_{(2)}}.(\mathrm{d}(\overline{d_{(1)}})\mathrm{d}(c_{(1)})) \otimes c_{(2)}\end{aligned} \tag{3.2.2}$$

and

$$\begin{aligned}&(\overline{\Delta} \otimes_{(\gamma,\alpha)} \Delta)(\overline{c} \otimes d)\\ &= \sum \overline{c_{(1)}} \otimes d_{(1)} \otimes (\overline{c_{(2)}}.\mathrm{d}(\overline{c_{(1)}})).\mathrm{d}(d_{(1)}) \otimes d_{(2)},\end{aligned}$$

so that

$$((\overline{\Delta} \otimes_{(\gamma,\alpha)} \Delta)(\overline{c} \otimes d))^*$$
$$= \sum \overline{d_{(1)}} \otimes c_{(1)} \otimes \overline{d_{(2)}}.\mathrm{d}(\overline{d_{(1)}}c_{(1)}) \otimes (\mathrm{d}(\overline{d_{(1)}}c_{(1)}).c_{(2)}).\mathrm{d}(\overline{d_{(1)}}c_{(1)})$$

which equals the right hand side of (3.2.2).◇

By regarding sesquilinear forms on a double-module coalgebra $\mathcal{C}$ as elements of $\mathbf{L}(\overline{\mathcal{C}} \otimes \mathcal{C})$, we obtain the notion of the convolution product of sesquilinear forms on double-module coalgebras.

A sesquilinear form L on a vector space is called positive if $L(v,v) \geq 0$ for all $v \in V$. For a coalgebra $\mathcal{C}$ and $b \in \mathcal{C}$ we denote by L^b the sesquilinear form on $\mathcal{C}$ with

$$L^b(c,d) = L(c - \epsilon(c)b, d - \epsilon(d)b).$$

A sesquilinear form A on $\mathcal{C}$ is called conditionally positive if

$$A(c,c) \geq 0 \qquad \text{for all } c \in \mathcal{C} \text{ with } \epsilon(c) = 0.$$

For a $\mathbb{Z}_2$-graded coalgebra the convolution exponentials $\exp_\star(tA)$ of an even sesquilinear form A on $\mathcal{C}$ are positive for all $t \geq 0$ if and only if A is hermitian and conditionally positive. The proof of this rests on the question under which additional conditions on the sesquilinear forms the convolution product of two positive sesquilinear forms again is positive. For $q = 1$ no extra condition is needed. For $\mathbb{L} = \mathbb{M} = \mathbb{Z}_2$ and $q = -1$ one of the sesquilinear forms has to be even. In the general case we have

3.2.2. PROPOSITION. *Let $\mathcal{C}$ be a double-module coalgebra. Then, under the condition that B is α-invariant or A is even, the convolution product $A \star B$ of two sesquilinear forms A and B is again positive.*

PROOF: We have for $c \in \mathcal{C}$

$$(A \star B)(c,c) = \sum_{i,j} A(\overline{c_{1i}} \otimes c_{1j}) B(\overline{c_{2i}}.\mathrm{d}(\overline{c_{1i}}c_{1j}) \otimes c_{2j}) \tag{3.2.3}$$

which, in the case when A is even equals

$$\sum_{i,j} A(\overline{c_{1i}} \otimes c_{1j}) B(\overline{c_{2i}} \otimes c_{2j})$$

which is ≥ 0 because the Schur product of two positive definite matrices is again positive. If B is α-invariant (3.2.3) equals

$$\sum_{i,j} A(\overline{c_{1i}} \otimes c_{1j}) B(\overline{c_{2i}}.\mathrm{d}(\overline{c_{1i}}) \otimes c_{2j}.\mathrm{d}(c_{1j})^{-1})$$
$$= \sum_{i,j} A(\overline{c_{1i}} \otimes c_{1j}) B(\overline{\mathrm{d}(c_{1i}).c_{2i}} \otimes \mathrm{d}(c_{1j}).c_{2j})$$

which is ≥ 0 for the same reason.◇

Now we reduce everything to the symmetric case. We equip $\mathcal{C}_\gamma$ with the trivial double-module structure. Then $\overline{(\mathcal{C}_\gamma)}$ is identified with $\overline{\mathcal{C}}_\gamma = \mathbb{CL} \otimes \overline{\mathcal{C}}$ via the linear map given by

$$\overline{\lambda \otimes c} \mapsto \lambda^* \otimes \overline{c}.\lambda^*$$

and $\overline{\mathcal{C}}_\gamma \otimes \mathcal{C}_\gamma$ is mapped into $(\overline{\mathcal{C}} \otimes \mathcal{C})_\gamma$ via the linear map Θ with

$$\Theta((\lambda \otimes \overline{c}) \otimes (\lambda' \otimes c')) = (\lambda\lambda') \otimes (\overline{c}.\lambda' \otimes c').$$

3.2.3. Proposition. *The map Θ is a coalgebra homomorphism from $(\overline{(\mathcal{C}_\gamma)} \otimes \mathcal{C}_\gamma, \overline{(\Delta_\gamma)} \otimes_1 \Delta_\gamma, \overline{(\epsilon_\gamma)} \otimes \epsilon_\gamma)$ to $((\overline{\mathcal{C}} \otimes_{(\gamma,\alpha)} \mathcal{C})_\gamma, (\overline{\Delta} \otimes_{(\gamma,\alpha)} \Delta)_\gamma, (\overline{\epsilon} \otimes \epsilon)_\gamma)$.*

Proof: We have

$$\begin{aligned}
&(\Theta \otimes \Theta) \circ (\overline{\Delta_{\mathbb{L}}} \otimes_1 \Delta_{\mathbb{L}})((\overline{\lambda \otimes c}) \otimes (\lambda' \otimes c')) \\
&= (\Theta \otimes \Theta)(\sum_{i,j} (\overline{\lambda \otimes c_{1i}}) \otimes (\lambda' \otimes c'_{1j}) \otimes (\overline{\lambda \mathrm{d}(c_{1i}) \otimes c_{2i}}) \otimes (\lambda' \mathrm{d}(c'_{1j}) \otimes c'_{2j})) \\
&= (\Theta \otimes \Theta) \sum (\lambda^* \otimes \overline{c_{(1)}}.\lambda^*) \otimes (\lambda' \otimes c_{(1)}{}') \\
&\qquad \otimes (\mathrm{d}(c_{(1)})^* \lambda^* \otimes \overline{c_{(2)}}.(\mathrm{d}(c_{(1)})^* \lambda^*)) \otimes (\lambda' \mathrm{d}(c_{(1)}{}') \otimes c_{(2)}{}') \\
&= \sum \lambda^* \lambda' \otimes \overline{c_{(1)}}.(\lambda^* \lambda') \otimes c_{(1)}{}' \\
&\qquad \otimes \mathrm{d}(c_{(1)})^* \lambda^* \lambda' \mathrm{d}(c_{(1)}{}') \otimes \overline{c_{(2)}}.(\mathrm{d}(c_{(1)})^* \lambda^* \lambda' \mathrm{d}(c_{(1)}{}')) \otimes c_{(2)}{}'
\end{aligned} \tag{3.2.4}$$

and

$$\begin{aligned}
&((\overline{\Delta} \otimes_{(\gamma,\alpha)} \Delta)_\gamma \circ \Theta)(\lambda^* \otimes \overline{c}.\lambda^*) \otimes (\lambda' \otimes c')) \\
&= (\overline{\Delta} \otimes_{(\gamma,\alpha)} \Delta)_\gamma (\lambda^* \lambda' \otimes \overline{c}.(\lambda^* \lambda') \otimes c') \\
&= \sum \lambda^* \lambda' \otimes (\overline{c_{(1)}}.(\lambda^* \lambda') \otimes c_{(1)}{}') \\
&\qquad \otimes \lambda^* \lambda' \mathrm{d}(\overline{c_{(1)}}.(\lambda^* \lambda') c_{(1)}{}') \otimes (\overline{c_{(2)}}.(\mathrm{d}(c_{(1)})^* \lambda^* \lambda' \mathrm{d}(c_{(1)}{}'))) \otimes c_{(2)}{}'
\end{aligned}$$

which is equal to (3.2.4).◇

3.2.4. Proposition. *The map*

$$(r_{\mathbb{L}} \circ \Theta)^* : (\overline{\mathcal{C}} \otimes_{(\gamma,\alpha)} \mathcal{C})^* \to (\overline{(\mathcal{C}_\gamma)} \otimes \mathcal{C}_\gamma)^*$$

is an algebra monomorphism. Moreover, for an α-invariant sesquilinear form A on $\mathcal{C}$ the sesquilinear form $(r_{\mathbb{L}} \circ \Theta)^*(A)$ on $\mathcal{C}_\gamma$ is hermitian (positive, conditionally positive) if and only if A is hermitian (positive, conditionally positive).

Proof: Since $r_{\mathbb{L}}$ and Θ are coalgebra homomorphisms $r_{\mathbb{L}} \circ \Theta$ is a coalgebra homomorphism and the first assertion follows. We have

$$A(\overline{c} \otimes c') = (A \circ r_{\mathbb{L}} \circ \Theta)((\overline{\mathrm{e} \otimes c}) \otimes (\mathrm{e} \otimes c'))$$

which shows that $(r_{\mathbb{L}} \circ \Theta)^*$ is injective. Now suppose that A is α-invariant. Then

$$\begin{aligned}(A^* \circ r_{\mathbb{L}} \circ \Theta)((\overline{\lambda \otimes c}) \otimes (\lambda' \otimes c')) &= A^*(\bar{c}.(\lambda^* \lambda') \otimes c') \\ &= A(\overline{c'} \otimes ((\lambda')^* \lambda).c) \\ &= A(\overline{\lambda'.c'} \otimes \lambda.c) \\ &= (A \circ r_{\mathbb{L}} \circ \Theta)^*((\overline{\lambda \otimes c}) \otimes (\lambda' \otimes c')).\end{aligned}$$

Moreover,

$$(A \circ r_{\mathbb{L}} \circ \Theta)((\overline{\lambda \otimes c}) \otimes \lambda \otimes c) = A(\bar{c} \otimes c)$$

which shows that $A \circ r_{\mathbb{L}} \circ \Theta$ is hermitian (positive, conditionally positive) if and only if A is.◇

The following result is now an immediate corollary to the preceding proposition and to Theorem 3.1 of [70].

3.2.5. **Theorem.** *Let A be an α-invariant sesquilinear form on a double-module coalgebra $\mathcal{C}$. We consider the following statements:*

(i) *A is conditionally positive*
(ii) *there is an element c in $\mathcal{C}$ with $\epsilon(c) = 1$ such that A^c is positive*
(iii) *A^c is positive for all $c \in \mathcal{C}$ with $\epsilon(c) = 1$*
(iv) *A is hermitian and conditionally positive*
(v) *$\exp_\star(tA)$ is positive for all $t \geq 0$.*

Then the implications

$$\begin{array}{ccccc} (i) & \Leftrightarrow & (ii) & \Leftrightarrow & (iii) \\ \Uparrow & & & & \\ (iv) & \Leftrightarrow & (v) & & \end{array}$$

hold.◇

Next we have the following proposition.

3.2.6. **Proposition.** *Let $\mathcal{B}$ be an involutive (γ, α)-bialgebra. Then the linear mapping*

$$K : \overline{\mathcal{B}} \otimes \mathcal{B} \to \mathcal{B}$$

given by

$$K(\bar{b} \otimes c) = b^* c$$

is an even coalgebra epimorphism. Moreover, K is α-invariant, and $K^*(\varphi)$ is hermitian (positive, conditionally positive) if and only if φ is hermitian (positive, conditionally positive).

Proof: We have

$$\begin{aligned}&((K \otimes K) \circ (\overline{\Delta} \otimes_{(\gamma,\alpha)} \Delta))(\bar{b} \otimes c) \\ &= (K \otimes K)(\sum (\overline{b_{(1)}} \otimes c_{(1)} \otimes \overline{b_{(2)}}.(\mathrm{d}(b_{(1)})^* \mathrm{d}(c_{(1)})) \otimes c_{(2)}) \\ &= \sum b_{(1)}{}^* c_{(1)} \otimes b_{(2)}{}^*.(\mathrm{d}(b_{(1)})^* \mathrm{d}(c_{(1)})) c_{(2)} \\ &= \Delta(b^* c) \\ &= (\Delta \circ K)(\bar{b} \otimes c)\end{aligned}$$

which shows that K is a coalgebra homomorphism. Since $K(\bar{e} \otimes b) = b$, the map K is surjective. For a linear functional φ on $\mathcal{B}$

$$K^*(\varphi)(\bar{b} \otimes c) = \varphi(b^* c)$$

and it is immediate that $K^*(\varphi)$ is hermitian (positive, conditionally positive) if and only if φ is.$\diamondsuit$

Now we have as a corollary to Theorem 3.2.5 and Prop. 3.2.6

3.2.7. THEOREM. *Let ψ be an α-invariant linear functional on an involutive (γ, α)-bialgebra $\mathcal{B}$. Then the following statements are equivalent:*

(i) *ψ is hermitian and conditionally positive*
(ii) *$\exp_\star(t\psi)$ is positive for all $t \geq 0$.*

Finally, the next theorem, together with Theorem 1.9.7, establishes the 1-1-correspondence between (equivalence classes of) white noises on $\mathcal{B}$ and certain linear functionals on $\mathcal{B}$.

3.2.8. THEOREM. *Let ψ be an α-invariant hermitian, conditionally positive linear functional on an involutive (γ, α)-bialgebra $\mathcal{B}$ and assume that ψ vanishes at* **1**. *Then there is a white noise on $\mathcal{B}$ with infinitesimal generator ψ.*$\diamondsuit$

In the sequel, by a *generator on $\mathcal{B}$* we mean an α-invariant hermitian, conditionally positive linear functional on $\mathcal{B}$ that vanishes at **1**.

3.3. Symmetrization of white noise. Involutive (γ, α)-bialgebras give rise to certain symmetric $*$-bialgebras, and a generator on an involutive (γ, α)-bialgebra determines a generator on its symmetrization. The relationship between the two white noises given by these generators is very easily described.

3.3.1. THEOREM. *Let ψ be a generator on an involutive (γ, α)-bialgebra $\mathcal{B}$. Denote by $\psi_{(\gamma,\alpha)}$ the generator $r_{\mathbb{L}}^*(\psi)$ on $\mathcal{B}_{(\gamma,\alpha)}$ associated with ψ, and assume that $(j_{st}^{(\gamma,\alpha)})_{(s,t) \in T_{\mathrm{L}}}$, $\mathrm{L} < \infty$, is a white noise on $\mathcal{B}_{(\gamma,\alpha)}$ with generator $\psi_{(\gamma,\alpha)}$. We put, for homogeneous $b \in \mathcal{B}$ and $0 \leq s \leq t \leq \mathrm{L}$,*

$$j_{st}(b) = j_{st}^{(\gamma,\alpha)}(b)\, j_{t\mathrm{L}}^{(\gamma,\alpha)}(\mathrm{d}_b).$$

Then $(j_{st})_{0 \leq s \leq t \leq \mathrm{L}}$ forms a white noise on $\mathcal{B}$ with generator ψ.

PROOF: We show that (j_{st}) satisfies the axioms (WN1)-(WN4). Using the fact that the increments $j_{st}^{(\gamma,\alpha)}$ and $j_{s't'}^{(\gamma,\alpha)}$ commute for disjoint intervals (s,t) and (s',t') and using $\Delta_\gamma \lambda = \lambda \otimes \lambda$, $\lambda \in \mathbb{L}$, we obtain for $r \leq s \leq t$ and $b \in \mathcal{B}$

$$\begin{aligned}
(j_{rs} \star j_{st})(b) &= \sum j_{rs}(b_{(1)}) j_{st}(b_{(2)}) \\
&= \sum j_{rs}^{(\gamma,\alpha)}(b_{(1)}) j_{s\mathrm{L}}^{(\gamma,\alpha)}(\mathrm{d}(b_{(1)})) j_{st}^{(\gamma,\alpha)}(b_{(2)}) j_{t\mathrm{L}}^{(\gamma,\alpha)}(\mathrm{d}(b_{(2)})) \\
&= \sum j_{rs}^{(\gamma,\alpha)}(b_{(1)}) j_{st}^{(\gamma,\alpha)}(\mathrm{d}(b_{(1)}) b_{(2)}) j_{t\mathrm{L}}^{(\gamma,\alpha)}(\mathrm{d}(b_{(1)}) \mathrm{d}(b_{(2)})) \\
&= j_{rt}^{(\gamma,\alpha)}(b) j_{t\mathrm{L}}^{(\gamma,\alpha)}(\mathrm{d}(b)) \\
&= j_{rt}(b)
\end{aligned}$$

which is the increment property (WN1). Next, for $b, b' \in \mathcal{B}$, $s \le t \le s' \le t'$,

$$\begin{aligned} j_{st}(b) j_{s't'}(b') &= j_{st}^{(\gamma,\alpha)}(b) j_{tL}^{(\gamma,\alpha)}(\mathrm{d}_b) j_{s't'}^{(\gamma,\alpha)}(b') j_{t'L}^{(\gamma,\alpha)}(\mathrm{d}'_b) \\ &= j_{s't'}^{(\gamma,\alpha)}(\mathrm{d}_b.b') j_{st}^{(\gamma,\alpha)}(b) j_{tL}^{(\gamma,\alpha)}(\mathrm{d}_b) j_{t'L}^{(\gamma,\alpha)}(\mathrm{d}_{b'}) \\ &= j_{s't'}^{(\gamma,\alpha)}(\mathrm{d}_b.b') j_{t'L}^{(\gamma,\alpha)}(\mathrm{d}_b \mathrm{d}_{b'} \mathrm{d}(b)^{-1}) j_{st}^{(\gamma,\alpha)}(b) j_{tL}^{(\gamma,\alpha)}(\mathrm{d}_b) \\ &= j_{s't'}(\mathrm{d}_b.b') j_{st}(b) \end{aligned}$$

where we also used the relations in $\mathcal{B}_\alpha$. Now let $t_1 < \cdots < t_{n+1}$ and $b_1, \ldots, b_n \in \mathcal{B}$. Then

$$\begin{aligned} &\Phi(j_{t_1 t_2}(b_1) \ldots j_{t_n t_{n+1}}(b_n)) \\ &= \Phi(j_{t_1 t_2}^{(\gamma,\alpha)}(b_1) j_{t_2 L}^{(\gamma,\alpha)}(\mathrm{d}(b_1)) \ldots j_{t_n t_{n+1}}^{(\gamma,\alpha)}(b_n) j_{t_{n+1} L}^{(\gamma,\alpha)}(\mathrm{d}(b_n))) \\ &= \exp_\star((t_2 - t_1)\psi_\alpha)(b_1) \exp_\star((t_3 - t_2)\psi_\alpha)(\mathrm{d}(b_1) b_2) \ldots \\ &\qquad \ldots \exp_\star((t_{n+1} - t_n)\psi_\alpha)(\mathrm{d}(b_1) \ldots \mathrm{d}(b_{n-1}) b_n) \\ &\qquad \exp_\star((\mathrm{L} - t_{n+1})\psi_\alpha)(\mathrm{d}(b_1) \ldots \mathrm{d}(b_{n-1}) \mathrm{d}(b_n)) \\ &= \varphi_{t_2 - t_1}(b_1) \ldots \varphi_{t_{n+1} - t_n}(b_n) \end{aligned}$$

which shows that (j_{st}) has the factorization property for time-ordered correlations. However, since ψ is assumed to be α-invariant, the white noise factorization property (1.3.1) holds. This gives the independence of increments (WN2). (WN3) and (WN4) follow by construction. This proves that (j_{st}) is a white noise on $\mathcal{B}$. Its infinitesimal generator is ψ.◇

The restriction $\mathrm{L} < \infty$ can be omitted, because $j_{t,\infty}^{(\gamma,\alpha)}(\lambda)$ will always make sense in the special realization of $(j_{st}^{(\gamma,\alpha)})$ which we will describe in the next chapter.

Suppose that ψ is a generator on the involutive (α,γ)-bialgebra $\mathcal{B}$. We form $\mathcal{B}^{\mathrm{op}}$ and $(\mathcal{B}^{\mathrm{op}})_{(\gamma,\alpha)}$, and ψ_α as a generator on $(\mathcal{B}^{\mathrm{op}})_{(\gamma,\alpha)}$. Let $h_{st}^{(\gamma,\alpha)}$ be a white noise on $(\mathcal{B}^{\mathrm{op}})_{(\gamma,\alpha)}$ with generator ψ_α. We put $j_{st}^{(\alpha,\gamma)} = (h_{st}^{(\gamma,\alpha)})^{\mathrm{op}} = h_{\mathrm{L}-t,\mathrm{L}-s}^{(\gamma,\alpha)}$. Then

$$j_{st}(b) = j_{0s}^{(\alpha,\gamma)}(\mathrm{d}_b)\, j_{st}^{(\alpha,\gamma)}(b) \tag{3.3.1}$$

gives a white noise on $\mathcal{B}$ with generator ψ.

There is a more general way to obtain a realization of (α,γ)-white noise. We know that $j_{st}^{(\alpha,\gamma)}$ is a (symmetric) white noise on $\mathcal{B}_{(\alpha,\gamma)} = ((\mathcal{B}^{\mathrm{op}})_{(\gamma,\alpha)})^{\mathrm{op}}$. Thus, if we consider ψ_α as a generator on $\mathcal{B}_{(\alpha,\gamma)}$, then for *any* realization $j_{st}^{(\alpha,\gamma)}$ of a white noise on $\mathcal{B}^{(\alpha,\gamma)}$ with generator ψ_α we have that (3.3.1) is a realization of a white noise on $\mathcal{B}$ with generator ψ.

4. White noise on Bose Fock space.

Using the symmetrization principle and the fact that group-like elements give rise to second quantization processes, we can immediately apply the results of Chapter 2 to the case of a general (γ, α)-bialgebra. Additive twisted white noise is an integral over the second quantization process of the action α with respect to a sum of creation, preservation, annihilation and scalar processes. After describing additive noise, we observe that, in complete analogy to the symmetric case, we can write a twisted white noise as the solution of a quantum stochastic differential equation against additive twisted white noise. Bose and Fermi quantum Brownian motion [22, 6, 7] and the 'quantum Poisson process' are special cases of additive white noise. The Azéma martingales [10, 26, 61] are closely related to additive q-white noise and to interpolations between Bose and Fermi quantum Brownian motion; for the latter see also [50].

In the case of multiplicative white noise, we obtain a particularly simple form of the quantum stochastic differential equation; it can be written as a linear quantum stochastic differential equation on the tensor product of the Fock space with some finite-dimensional 'initial space'. We obtain a characterization of unitary white noises which turn out to be unitary evolutions in the sense of [41, 42] on $\mathbb{C}^n \otimes \mathcal{F}(L^2(\mathbb{R}_+, H))$ where H can be an infinite-dimensional Hilbert space.

In the case of a $\mathbb{Z}_2$-graded cocommutative $*$-bialgebra white noise can be written down explicitly without using kernels or quantum stochastic calculus. This generalizes the results for group Hopf algebras and tensor Hopf algebras. We obtain a classification of the infinitely divisible representations [56, 86] of $\mathbb{Z}_2$-graded Lie $*$-algebras.

4.1. Group-like elements and realization of white noise. Since the elements of $\mathbb{L} \subset \mathcal{B}_{(\gamma,\alpha)}$ are group-like, in view of Theorem 3.3.1 we must investigate the behaviour of $j_{st}(b)$ for $b \in \mathcal{B}$, b group-like, $\mathcal{B}$ a symmetric $*$-bialgebra. Before we begin with a combinatorical lemma, we introduce some notation. In a similar way, it was already used in Chapter 2. For a vector space V and $\omega = \{t_1 < \cdots < t_n\} \in \mathcal{S}_n$, we put

$$V^{\otimes\omega} = V_{t_1} \otimes \ldots \otimes V_{t_n}$$

with V_{t_k} a copy of V and we put $V^{\otimes\emptyset} = \mathbb{C}$. For a linear mapping $R : V \rightarrow W$ from a vector space V to a vector spave W we let $R^{\otimes\omega}$ denote the mapping

$$R_{t_1} \otimes \ldots \otimes R_{t_n} : V^{\otimes\omega} \rightarrow W^{\otimes\omega}$$

with R_{t_k} a copy of R and we put $R^{\otimes\emptyset} = \mathrm{id}$. Furthermore, for linear mappings $R_{t_1}, \ldots, R_{t_m} : V \rightarrow W$ and $\tau = \{t_1 < \cdots < t_m\} \subset \omega$ we denote by

$$(R_{t_1} \otimes \ldots \otimes R_{t_m})^{(\omega)} : V^{\otimes\omega} \rightarrow W^{\otimes\omega}$$

the ampliation of $R_{t_1} \otimes \ldots \otimes R_{t_m}$ to $V^{\otimes\omega}$.

4.1.1. LEMMA. *Let V be a vector space, $\mathcal{C}$ a coalgebra and $\rho : \mathcal{C} \rightarrow \mathbf{L}(V)$ linear. Then the equality*

$$(\rho^{\otimes\omega} \circ \Delta_\omega)(b) = \sum_{\tau \subset \omega} ((\rho - \epsilon\,\mathrm{id})^{\otimes\tau} \circ \Delta_\tau)(b)^{(\omega)} \tag{4.1.1}$$

holds for all $\omega \in \mathcal{S}$.

PROOF: (by induction on $n = \#\omega$) Equality (4.1.1) is correct for $n = 0$. Suppose it holds for n. Using the coassociativity and the counit property, we obtain for $\omega = \sigma \cup \{t\}$, $\#\sigma = n$, $t > \hat{\sigma}$, that the right hand side of (4.1.1) equals

$$\begin{aligned}
&\sum_{\tau \subset \sigma} \Big(((\rho - \epsilon\,\mathrm{id})^{\otimes\tau} \circ \Delta_\tau(b)^{(\sigma)}) \otimes \mathrm{id} + ((\rho - \epsilon\,\mathrm{id})^{\otimes\tau} \otimes \rho) \circ \Delta_{\tau\cup\{t\}}(b)^{(\omega)} \\
&\qquad - ((\rho - \epsilon\,\mathrm{id})^{\otimes\tau} \otimes \epsilon\,\mathrm{id}) \circ \Delta_{\tau\cup\{t\}}(b)^{(\omega)} \Big) \\
&= (\rho^{\otimes\sigma} \circ \Delta_\sigma(b)) \otimes \mathrm{id} + ((\rho^{\otimes\sigma} \circ \Delta_\sigma) \otimes \rho) \circ \Delta(b) \\
&\qquad - ((\rho^{\otimes\sigma} \circ \Delta_\sigma) \otimes \epsilon\,\mathrm{id}) \circ \Delta(b) \\
&= (\rho^{\otimes\sigma} \otimes \rho) \circ (\Delta_\sigma \otimes \mathrm{id}) \circ \Delta(b)
\end{aligned}$$

which is the left hand side of (4.1.1).◇

Let D be a pre-Hilbert space and $R \in \mathbf{L}(D)$. We define the *second quantization process* of R to be the family $(\Gamma_{st}(R))_{(s,t)\in T}$ of linear operators on $\mathcal{E}_D$ given by the kernels

$$\gamma_{st}(\sigma, \tau, \varrho) = \begin{cases} \prod_{r\in\tau} \lambda(r)(R - \mathrm{id}) & \text{if } \sigma = \varrho = \emptyset \text{ and } \tau \subset [s,t) \\ 0 & \text{otherwise.} \end{cases}$$

Then $\gamma_{st} \in \mathcal{A}_D$ and, if $R \in \mathbf{H}(D)$, we have that $\gamma_{st} \in \mathcal{A}_{D,\mathrm{h}}$.

4.1.2. PROPOSITION. *Let $\mathcal{C}$ be a coalgebra, let $b \in \mathcal{C}$ be group-like and let $(\eta, \rho, \vartheta, \psi)$ be as in Section 2.3. Then*

$$\begin{aligned} k_{st}^{\eta,\rho,\vartheta,\psi}(b)(\sigma,\tau,\varrho) &= \exp((t-s)\psi(b)) \prod_{r\in\sigma} a^*(r)(\eta(b)) \\ &\qquad \prod_{r\in\tau} \lambda(r)(\rho(b) - \mathrm{id}) \prod_{r\in\rho} a(r)(\vartheta(b)) \end{aligned} \tag{4.1.2}$$

for $\sigma \cup \tau \cup \varrho \subset [s,t)$, $t < \mathrm{L}$. Moreover, $j_{st}(b) = \underline{k}_{st}(b)$ sends an exponential vector $\mathrm{E}(f)$, $f \in L^2(\mathbb{R}_+) \otimes D$, to the multiple

$$\begin{aligned} &\exp((t-s)\psi(b) + \int_s^t \langle \vartheta(b), f(r)\rangle \mathrm{d}r) \\ &\qquad \mathrm{E}(f\chi_{[0,s)\cup[t,\infty)} + (\rho(b)f + \eta(b))\chi_{[s,t)}) \end{aligned} \tag{4.1.3}$$

of an exponential vector. Furthermore, $j_{st}(b)$ is the unique solution in $\underline{\mathcal{A}}_D$ of the quantum stochastic integral equation

$$j_{st}(b) = \mathrm{id} + \int_s^t j_{sr}(b)(\mathrm{d}A_r^*(\eta(b)) + \mathrm{d}\Lambda_r(\rho(b) - \mathrm{id}) + \mathrm{d}A_r(\vartheta(b)) + \psi(b)\mathrm{d}r) \tag{4.1.4}$$

In particular, if $\eta(b) = \vartheta(b) = 0$ and $\psi(b) = 0$ the process $j_{st}(b)$ equals the second quantization process $\Gamma_{st}(\rho(b))$ of $\rho(b)$.

PROOF: Since b is group-like we have $\Delta_n b = b^{\otimes n}$ and formula (4.1.2) and equation (4.1.4) follow directly from the theory developed in Chapter 2.

For an exponential vector $\mathrm{E}(f) \in \mathcal{E}_D$ and $\omega = \{t_1 < \cdots < t_n\}$ with $s \leq t_1, t_n < t$, we compute

$$\begin{aligned} & k_{st}(\sigma, \tau, \varrho)\mathrm{E}(f)(\omega) \\ & = \exp((t-s)\psi(b) + \int_s^t \langle \vartheta(b), f(r)\rangle \mathrm{d}r) \\ & \qquad \sum_{\sigma \dot\cup \tau \dot\cup \epsilon = \omega} (\prod_{r\in\sigma} a^*(r)\eta(b))(\prod_{r\in\tau} \lambda(r)(\rho(b) - \mathrm{id})) \otimes^{\rightarrow}_{r\in\tau\cup\epsilon} f(r) \end{aligned}$$

which gives (4.1.3) by an application of Lemma 4.1.1.◇

In particular, $\Gamma_{st}(R)$ is the unique solution of

$$\Gamma_{st}(R) = \mathrm{id} + \int_s^t \Gamma_{sr}(R)\mathrm{d}\Lambda_r(R - \mathrm{id})$$

and

$$(\Gamma_{st}(R)F)(\omega) = R^{\otimes\omega} F(\omega) \text{ for } \omega \subset [s, t).$$

The structure of white noise on $*$-semi-group bialgebras $\mathbb{C}G$ and, therefore, on group $*$-Hopf algebras is now clear. The operators $j_{st}(x)$, $x \in G$, are given by kernel processes of the form (4.1.2) with $\vartheta = \tilde\eta$ and they map $\mathrm{E}(f)$ to (4.1.3). In the case of a group G, the $j_{st}(x)$ are unitary and this is the well-known formula for $j_{st}(x)\mathrm{E}(f)$ that can be found in [32, 63]. However, the applications of Prop. 4.1.2 are much wider.

4.1.3. PROPOSITION. *Let ψ be a generator on the involutive (γ,α)-bialgebra $\mathcal{B}$ and let $j_{st}^{(\gamma,\alpha)}$ be the realization on Bose Fock space of the white noise on $\mathcal{B}_{(\gamma,\alpha)}$ with generator ψ_α. Then for $\lambda \in \mathbb{L}$ the operators $j_{st}^{(\gamma,\alpha)}(\lambda)$ form the second quantization process $\Gamma_{st}(\rho_\alpha(\lambda))$ of $\rho_\alpha(\lambda)$. Moreover, we have for $b \in \mathcal{B}$ and $\lambda \in \mathbb{L}$*

$$\rho_\alpha(\lambda)\eta(b) = \eta(\lambda.b).$$

PROOF: Since λ is group-like and $\eta_\alpha(\lambda) = \eta_\alpha(\lambda^*) = \psi_\alpha(\lambda) = 0$ we deduce from Prop. 4.1.2 that

$$j_{st}^{(\gamma,\alpha)}(\lambda) = \Gamma_{st}(\rho_\alpha(\lambda)).$$

Next we have for $b \in \operatorname{Kern}\epsilon$ and $\lambda \in \mathbb{L}$

$$\begin{aligned} \|\eta_\alpha(\lambda.b) - \eta_\alpha(\lambda b)\|^2 &= \psi_\alpha((\lambda.b - \lambda b)^*(\lambda.b - \lambda b)) \\ &= \psi_\alpha((b^*.\lambda^*)(\lambda.b)) + \psi_\alpha(\lambda^*(b^*.\lambda^*)(\lambda.b)\lambda) \\ &\qquad - \psi_\alpha((b^*.\lambda^*)(\lambda.b)\lambda) - \psi_\alpha(\lambda^*(b^*.\lambda^*)(\lambda.b)) \\ &= 0. \end{aligned}$$

where we used the relations in $\mathcal{B}_{(\gamma,\alpha)}$ and the α-invariance of ψ.◇

In the following, we denote the $*$-representation $\lambda \mapsto \rho_\alpha(\lambda)$ of $\mathbb{L}$ on D again by the symbol α, so that $\eta(\lambda.b) = \alpha_\lambda\eta(b)$.

A remark on the commutation factor case under the assumption that $\mathsf{M} = G^{\mathrm{inv}}$ for some group G: Since ψ is M-even, we have that D is the orthogonal sum

$$D = \bigoplus_{\mu \in \mathsf{M}} D^{(\mu)}$$

of the linear subspaces $D^{(\mu)} = \eta(\mathcal{B}^{(\mu)})$ and $\alpha_\lambda \lceil \mathcal{B}^{(\mu)}$ is simply multiplication by $q(\lambda, \mu)$.

Now we are in a position to formulate the representation theorem for general (γ, α)-white noise.

4.1.4. THEOREM. *Let ψ be a generator on the involutive (γ, α)-bialgebra $\mathcal{B}$. Then the quantum stochastic backward integral equations*

$$j_{st}(b) = \epsilon(b)\mathrm{id} + \int_s^t (\mathrm{d}I_r \star j_{rt})(b) \tag{4.1.5}$$

with

$$\mathrm{d}I_r(b) = \mathrm{d}\tilde{I}_r(b)\Gamma_{r,\infty}(\alpha_{\mathrm{d}(b)}).$$

and

$$\tilde{I}_t = A_t^*(\eta(b)) + \Lambda_t(\rho(b) - \varepsilon(b)\mathrm{id}) + A_t(\eta(b^*)) + \psi(b)t$$

have unique solutions $j_{st}(b)$ in $\underline{\mathcal{A}}_D$. In the vacuum state, the j_{st} form a white noise on $\mathcal{B}$ with generator ψ.

Let $j_{st}^{(\gamma,\alpha)}(b)$ be the unique solution in $\underline{\mathcal{A}}_D$ of the quantum stochastic forward integral equation

$$\begin{aligned} j_{st}^{(\gamma,\alpha)}&(b) \\ &= \epsilon(b)\mathrm{id} + \int_s^t \sum (j_{sr}^{(\gamma,\alpha)}(b_{(1)}) \\ &\qquad (\mathrm{d}A_r^*(\eta(\mathrm{d}(b_{(1)}).b_{(2)})) + \mathrm{d}\Lambda_r(\alpha(\mathrm{d}(b_{(1)}))\rho(b_{(2)}) - \epsilon(b_{(2)})\mathrm{id}) \\ &\qquad + \mathrm{d}A_r(\eta(b_{(2)}^*)) + \psi(b_{(2)})\mathrm{d}r). \end{aligned} \tag{4.1.6}$$

Then

$$j_{st}(b) = j_{st}^{(\gamma,\alpha)}(b)\Gamma_{t,\infty}(\alpha_{\mathrm{d}(b)}). \tag{4.1.7}$$

PROOF: The realization $j_{st}^{(\gamma,\alpha)}(b)$ on Bose Fock space of a symmetric white noise on $\mathcal{B}_{(\gamma,\alpha)}$ with generator ψ_α is the unique solution in $\underline{\mathcal{A}}_D$ of the equation

$$j_{st}^{(\gamma,\alpha)}(b) = \epsilon(b)\mathrm{id} + \int_s^t \sum (\mathrm{d}\tilde{I}_r(b_{(1)}) j_{rt}^{(\gamma,\alpha)}(\mathrm{d}(b_{(1)})b_{(2)}). \tag{4.1.8}$$

Using the homomorphism property of ρ_α and the facts that η_α is a 1-cocycle with respect to ρ_α and ψ is α-invariant, we see that $j_{st}^{(\gamma,\alpha)}(b)$ is also the unique solution in $\underline{\mathcal{A}}_D$ of (4.1.6). By Theorem 3.3.1 and Prop. 4.1.3 we have that (4.1.7) is a realization of white noise with generator ψ. We multiply equation (4.1.8) by $\Gamma_{t,\infty}(\alpha_{\mathrm{d}(b)})$ and use

$$j_{st}^{(\gamma,\alpha)}(\mathrm{d}(b_{(1)})b_{(2)})\Gamma_{t,\infty}(\alpha_{\mathrm{d}(b)}) = \Gamma_{s,\infty}(\alpha_{\mathrm{d}(b_{(1)})})j_{st}(b_{(2)})$$

to obtain (4.1.5). The uniqueness assertion in the first part of the theorem follows from the fact that, if $\tilde{j}_{st}(b)$ is a solution of (4.1.5), then $\tilde{j}_{st}(b)\Gamma_{t,\infty}(\alpha^{-1}_{\mathrm{d}(b)})$ is a solution of (4.1.8). The latter has a unique solution in $\underline{A}_D$ by Theorem 2.5.3.◇

There is a version of Theorem 4.1.4 for (α, γ)-white noise. Equation (4.1.5) is replaced by the forward equation

$$j_{st}(b) = \epsilon(b)\mathrm{id} + \int_s^t (j_{sr} \star \mathrm{d}I_r)(b) \tag{4.1.9}$$

with

$$\mathrm{d}I_r(b) = \Gamma_r(\alpha_{\mathrm{d}(b)})\mathrm{d}\tilde{I}_r(b),$$

(4.1.6) is replaced by the backward equation

$$\begin{aligned} &j_{st}^{(\alpha,\gamma)}(b) \\ &= \epsilon(b)\mathrm{id} + \int_s^t \sum (\mathrm{d}A_r^*(\eta(b_{(1)})) + \mathrm{d}\Lambda_r(\rho(b_{(1)})\alpha_{\mathrm{d}(b_{(2)})} - \epsilon(b_{(1)})\mathrm{id}) \\ &\qquad + \mathrm{d}A_r(\eta(\mathrm{d}(b_{(2)})^* . b_{(1)}^*)) + \psi(b_{(1)})\mathrm{d}r) j_{rt}^{(\alpha,\gamma)}(b_{(2)}) \end{aligned} \tag{4.1.10}$$

and (4.1.7) is replaced by

$$j_{st}(b) = \Gamma_s(\alpha_{\mathrm{d}(b)}) j_{st}^{(\alpha,\gamma)}(b).$$

4.2. Primitive elements and additive white noise. We apply Theorem 4.1.4 to tensor-∗-Hopf algebras and show that the equations (4.1.5), (4.1.6) and (4.1.9), (4.1.10) can be rewritten as simpler integral equations with integrals against additive left white noise and additive right white noise respectively. These simpler equations are the precise analogues of the symmetric case. Again we start from a general consideration.

4.2.1. PROPOSITION. *Let $\mathcal{B}$ be a symmetric bialgebra, let $y \in \mathcal{B}$ be group-like and $b \in \mathcal{B}$ be $(1, y)$-primitive, and let $(\eta, \rho, \vartheta, \psi)$ be as in Theorem 2.3.3. Then we have for $\sigma\dot{\cup}\tau\dot{\cup}\varrho = \{t_1 < \cdots < t_n\}$*

$$\begin{aligned} &k_{st}^{(\eta,\rho,\vartheta,\psi)}(b)(\sigma, \tau, \varrho) \\ &= (\psi(b)\psi(y)^{-1}(\exp((t_1 - s)\psi(y)) - 1)\kappa(t_1; \sigma, \tau, \varrho)(y) + \kappa(t_1; \sigma, \tau, \varrho)(b)) \\ &\qquad \exp((t - t_1)\psi(y))(\kappa(t_2; \sigma, \tau, \varrho)(y) \ldots \kappa(t_n; \sigma, \tau, \varrho)(y)). \end{aligned} \tag{4.2.1}$$

Moreover, for $j_{st} = \underline{k}_{st}$ we have that

$$j_{st}(b) = \int_s^t \mathrm{d}I_r(b) j_{rt}(y), \tag{4.2.2}$$

and $j_{st}(b)$ is the unique solution in $\underline{A}_D$ of

$$j_{st}(b) - I_{st}(b) = \int_s^t j_{sr}(b)\mathrm{d}I_r(y). \tag{4.2.3}$$

PROOF: Equations (4.2.2) and (4.2.3) follow directly from the backward version of Theorem 2.5.1 because $\Delta b = b \otimes y + 1 \otimes b$ and $I_r(1) = 0$, $j_{sr}(1) = \text{id}$. We have

$$\Delta_n b = b \otimes y^{\otimes(n-1)} + 1 \otimes b \otimes y^{\otimes(n-2)} + \cdots + 1^{\otimes(n-1)} \otimes b. \tag{4.2.4}$$

Since $\kappa(t;\sigma,\tau,\varrho)(1) = 0$ only the first two summands of the right hand side of (4.2.4) survive, so that

$$\begin{aligned} & k_{st}(b)(\sigma,\tau,\varrho) \\ &= \varphi_{t_1-s}(b)\varphi_{t-t_1}(y)\kappa(t_1;\sigma,\tau,\varrho)(y)\dots\kappa(t_n;\sigma,\tau,\varrho)(y) \\ &\quad + \varphi_{t-t_1}(y)\kappa(t_1;\sigma,\tau,\varrho)(b)\kappa(t_2;\sigma,\tau,\varrho)(y)\dots\kappa(t_n;\sigma,\tau,\varrho)(y) \end{aligned}$$

which equals the right hand side of (4.2.1).◇

The structure of additive (γ,α)-white noise can now be described. Let V be a double-module $*$-vector space, i.e. V is an $\mathbb{L}$-graded involutive vector space and also a $\mathbb{CL}$-module such that $\text{d}(\lambda.v) = \lambda\text{d}(v)\lambda^{-1}$ and $(\lambda.v)^* = v^*\lambda^*$. Then a homogeneous $v \in V \subset \mathcal{T}(V)_{(\gamma,\alpha)}$ is $(1,\text{d}(v))$-primitive and, in combining Theorem 4.1.4 with Prop. 4.2.1, we have the following result.

4.2.2. THEOREM. *Let ψ be a generator on the involutive (γ,α)-Hopf algebra $\mathcal{T}(V)$. We put*

$$F_t(v) = \int_0^t \text{d}\tilde{I}_r(v)\Gamma_{r,\infty}(\alpha_{\text{d}(v)}) \tag{4.2.5}$$

for $v \in V$ homogeneous. Then, in the vacuum state, the F_t, $t \geq 0$, form an additive (γ,α)-white noise over V with generator ψ.

Let $L_t(v)$ be the unique solution in $\underline{A}_D$ of

$$L_t(v) - \tilde{I}_t(v) = \int_0^t L_r(v)\text{d}\Lambda_r(\alpha_{\text{d}(v)} - \text{id}). \tag{4.2.6}$$

Then

$$F_t(v) = L_t(v)\Gamma_{t,\infty}(\alpha_{\text{d}(v)}).\diamondsuit$$

The equations for a (α,γ)-Hopf algebra $\mathcal{T}(V)$ are

$$\check{F}_t(v) = \int_0^t \Gamma_r(\alpha_{\text{d}(v)})\text{d}\tilde{I}_r(v) \tag{4.2.7}$$

and

$$\check{L}_t(v) - \tilde{I}_t(v) = \check{F}_t(v) - \tilde{I}_t(v) = \int_0^t \text{d}\Lambda_r(\alpha_{\text{d}(v)} - \text{id})\check{F}_r(v).$$

We look at the equation for $\text{d}_v^{-1}v$. We have

$$\Delta_{(\gamma,\alpha)}(\text{d}_v^{-1}v) = (\text{d}_v^{-1}v) \otimes 1 + \text{d}_v^{-1} \otimes (\text{d}_v^{-1}v)$$

and

$$j_t^{(\gamma,\alpha)}(\mathrm{d}_v^{-1}v)$$
$$= \int_0^t \Gamma_r(\alpha_{\mathrm{d}(v)}^{-1})(\mathrm{d}A_r^*(\alpha_{\mathrm{d}(v)}^{-1}\eta(v)) + \mathrm{d}\Lambda_r(\alpha_{\mathrm{d}(v)}^{-1}\rho(v)) + \mathrm{d}A_r(\eta(v^*)) + \psi(v)\mathrm{d}r). \tag{4.2.8}$$

Thus $\Gamma_t(\mathrm{d}_v^{-1})L_t(v)$ equals the above integral and $F_t(v)$ is equal to this integral multiplied by $\Gamma_{0\infty}(\alpha_{\mathrm{d}(v)})$ from the left.

There is a remarkable consequence of the following

4.2.3. PROPOSITION. *Let $\mathcal{B}$ be a symmetric $*$-bialgebra and let $b \in \mathcal{B}$ be such that $b \otimes \mathbf{1}$ commutes with Δb. Then for any white noise (j_{st}) on $\mathcal{B}$ the family $(j_t(b))_{t\geq 0}$ is a family of commuting operators.*

PROOF: We have for $s \leq t$

$$\begin{aligned} j_s(b)j_t(b) &= \sum j_s(b)j_s(b_{(1)})j_{st}(b_{(2)}) \\ &= \sum j_s(bb_{(1)})j_{st}(b_{(2)}) \\ &= (M \circ (j_s \otimes j_{st}))(b \otimes \mathbf{1})(\Delta b) \\ &= (M \circ (j_s \otimes j_{st}))(\Delta b)(b \otimes \mathbf{1}) \\ &= \sum j_s(b_{(1)})j_{st}(b_{(2)})j_s(b) \\ &= j_t(b)j_s(b) \end{aligned}$$

where we only used the increment property, the condition $[b \otimes \mathbf{1}, \Delta b] = 0$, and the independence of increments.$\diamondsuit$

Applying this to $\mathcal{T}(V)_{(\gamma,\alpha)}$, we see that $L_t(v)$ form a *commutative process* for each $v \in V$. In the next section we will see that this is not true in general for $\check{L}_t(v)$ which gives an other justification of why we prefer left white noise.

We give a closer decription of the generators on $\mathcal{T}(V)$.

4.2.4. PROPOSITION. *Let D be a pre-Hilbert space and let*

$$\begin{aligned} \rho_0 &: V \to \mathbf{H}(D) \\ \eta_0 &: V \to D \\ \psi_0 &: V \to \mathbb{C} \\ \alpha &: \mathbb{L} \to \mathbf{H}(D) \end{aligned}$$

be such that ρ_0 and ψ_0 are hermitian, α is a $*$-representation of $\mathbb{L}$ on D,

$$D = \{\rho_0(v_1)\ldots\rho_0(v_k)\eta(v_{k+1}) : k \in \mathbb{N}, v_1, \ldots, v_{k+1} \in V\}$$

and

$$\rho_0(\lambda.v) = \alpha(\lambda)\rho_0(v)\alpha(\lambda)^{-1} \tag{4.2.9}$$
$$\eta_0(\lambda.v) = \alpha(\lambda)\eta_0(v) \tag{4.2.10}$$
$$\psi_0(\lambda.v) = \psi_0(v) \tag{4.2.11}$$

for all $v \in V$ and $\lambda \in \mathbb{L}$. Then, by recurrence, the equations

$$\rho\lceil V = \rho_0; \ \rho(bc) = \rho(b)\rho(c) \tag{4.2.12}$$
$$\eta\lceil V = \eta_0; \ \eta(bc) = \rho(b)\eta(c) + \eta(b)\varepsilon(c) \tag{4.2.13}$$
$$\psi(1) = 0; \ \psi\lceil V = \psi_0; \ \psi(bc) = \langle \eta(b^*), \eta(c)\rangle + \psi(b)\varepsilon(c) + \varepsilon(b)\psi(c) \tag{4.2.14}$$

define a generator ψ on $\mathcal{T}(V)$ such that $(\rho, \eta, \alpha) \sim (\rho_\psi, \eta_\psi, \alpha_\psi)$.

PROOF: Since $\mathcal{T}(V)$ is free, the procedure (4.2.12)-(4.2.14) defines a linear functional ψ on $\mathcal{T}(V)$. For $b \in \operatorname{Kern}\epsilon$ we have that $\psi(b^*b) = \|\eta(b)\|^2 \geq 0$ and ψ is conditionally positive. By assumption ψ_0 is hermitian, so that $\psi(v^*) = \overline{\psi(v)}$. For monomials M of length ≥ 2 we have $\psi(M^*) = \overline{\psi(M)}$ by (4.2.14). It follows that ψ is hermitian. The α-invariance of ψ_0 implies $\psi(\lambda.v) = \psi(v)$. Furthermore, using (4.2.9) and the defining equation (4.2.12) for ρ, we obtain

$$\rho(\lambda.b) = \alpha(\lambda)\rho(b)\alpha(\lambda)^{-1} \text{ for all } \lambda \in \mathbb{L}, b \in \mathcal{T}(V),$$

by induction on the length of monomials in $\mathcal{T}(V)$. Using this, (4.2.10) and (4.2.13) yield

$$\eta(\lambda.b) = \alpha(\lambda)\eta(b) \text{ for all } \lambda \in \mathbb{L}, b \in \mathcal{T}(V),$$

and we have for $b, c \in \operatorname{Kern}\epsilon$

$$\begin{aligned}\psi(\lambda.(bc)) &= \langle \eta(b^*.\lambda^*), \eta(\lambda.c)\rangle \\ &= \langle \eta(b^*), \alpha(\lambda)^{-1}\alpha(\lambda)\eta(c)\rangle \\ &= \langle \eta(b^*), \eta(c)\rangle \\ &= \psi(bc)\end{aligned}$$

which proves the α-invariance of ψ. The mappping $\eta_\psi(b) \mapsto \eta(b)$ extends to a unitary operator $\mathcal{U}$ from the completion of D_ψ to the completion of D such that

$$\begin{aligned}\eta_\psi &= \eta \circ \mathcal{U} \\ \rho_\psi(b) &= \mathcal{U}^* \rho(b)\mathcal{U} \\ \alpha_\psi(\lambda) &= \mathcal{U}^*\alpha(\lambda)\mathcal{U}.\diamond\end{aligned}$$

We have neglected one important aspect. Let ψ be a generator on the involutive (γ, α)-bialgebra $\mathcal{B}$. In particular, $\mathcal{B}$ is a double-module $*$-vector space. The triplet $(\rho_{\mathrm{a},0}, \eta_{\mathrm{a},0}, \psi_{\mathrm{a},0})$ with $\rho_{\mathrm{a},0} = \rho_\psi - \epsilon\,\mathrm{id}$, $\eta_{\mathrm{a},0} = \eta_\psi$ and $\psi_{\mathrm{a},0} = \psi$ is a triplet as it appears in Prop. 4.2.4 where $\alpha(\lambda)$ is given by $\eta_\psi(b) \mapsto \eta_\psi(\alpha_\lambda b)$. Hence there is an *additive* left white noise on the involutive vector space $\mathcal{B}$ given by the triplet $(\rho_{\mathrm{a},0}, \eta_{\mathrm{a},0}, \psi_{\mathrm{a},0})$. In other words, given a left white noise j_{st} on $\mathcal{B}$ we can associate with it an additive left white noise F_t on $\mathcal{B}$. Similarly, we associate an additive right white noise with a right white noise. Loooking at equations (4.1.5) and (4.2.5) and equations (4.1.9) and (4.2.7), we see that (4.1.5) can be written

$$j_{st}(b) = \epsilon(b)\mathrm{id} + \int_s^t (\mathrm{d}F_\tau \star j_{\tau t})(b)$$

and (4.1.9) can be written

$$j_{st}(b) = \epsilon(b)\mathrm{id} + \int_s^t (j_{sr} \star \mathrm{d}F_r)(b).$$

We also want simpler forward and backward equations for left and for right white noise respectively. Here we are facing the problem that left white noise is not forward adapted and right white noise is not backward adapted. Therefore, we multiply $j_{st}(b)$ by $\Gamma_{0\infty}(\alpha_{\mathrm{d}(b)})^{-1}$ from the left to obtain a forward adapted process $\tilde{j}_{st}$. This process satisfies

$$\begin{aligned}
&\tilde{j}_{st}(b)\\
&= \epsilon(b)\mathrm{id} + \int_s^t (\sum \Gamma_s(\alpha_{\mathrm{d}(b)}^{-1}) j_{sr}^{(\gamma,\alpha)}(\mathrm{d}_b^{-1} b_{(1)})\\
&\qquad (\mathrm{d}A_r^*(\eta(\mathrm{d}_b^{-1}\mathrm{d}(b_{(1)})b_{(2)})) + \mathrm{d}\Lambda_r(\alpha_{\mathrm{d}(b)}^{-1}\alpha_{\mathrm{d}(b_{(1)})}\rho(b_{(2)}) - \epsilon(b_{(2)})\mathrm{id})\\
&\qquad + \mathrm{d}A_r(\eta(b_{(2)}^*)) + \psi(b_{(2)})\mathrm{d}r))\\
&= \epsilon(b)\mathrm{id} + \int_s^t (\sum \tilde{j}_{sr}(\mathrm{d}(b_{(2)})^{-1}.b_{(1)})\Gamma_r(\alpha_{\mathrm{d}(b_{(2)})}^{-1})\\
&\qquad (\mathrm{d}A_r^*(\eta(\alpha_{\mathrm{d}(b_{(2)})}^{-1} b_{(2)})) + \mathrm{d}\Lambda_r(\alpha_{\mathrm{d}(b_{(2)})}^{-1}(\rho - \epsilon\,\mathrm{id})(b_{(2)}))\\
&\qquad + \mathrm{d}A_r(\eta(b_{(2)}^*)) + \psi(b_{(2)})\mathrm{d}r)).
\end{aligned}$$

If we multiply by $\Gamma_{0\infty}(\alpha_{\mathrm{d}(b)})$ from the left we obtain

$$\begin{aligned}
j_{st}(b) = \epsilon(b)\mathrm{id} + \int_s^t (\sum j_{sr}(b_{(1)})\Gamma_{0\infty}(\alpha_{\mathrm{d}(b_{(2)})})\Gamma_r(\alpha_{\mathrm{d}(b_{(2)})}^{-1})\\
(\mathrm{d}A_r^*(\eta(\alpha_{\mathrm{d}(b_{(2)})}^{-1} b_{(2)})) + \mathrm{d}\Lambda_r(\alpha_{\mathrm{d}(b_{(2)})}^{-1}(\rho - \epsilon\,\mathrm{id})(b_{(2)}))\\
+ \mathrm{d}A_r(\eta(b_{(2)}^*)) + \psi(b_{(2)})\mathrm{d}r))
\end{aligned}$$

where the integrals are to be understood in the obvious way. Comparing this with (4.2.8), we have

$$j_{st}(b) = \epsilon(b)\mathrm{id} + \int_s^t (j_{sr} \star \mathrm{d}F_r)(b)$$

which is in complete analogy to the symmetric case. In a similar way, we may write

$$j_{st}(b) = \epsilon(b)\mathrm{id} + \int_s^t (\mathrm{d}F_r \star j_{rt})(b)$$

for right white noise. The relationship between j_{st} and F_t goes further. We will see that F_t can be approximated by sums of the form $\sum_{0<t_{l+1}\leq t}(j_{t_l t_{l+1}} - \epsilon\,\mathrm{id})$ if the partition $\{0 = t_1 < \cdots < t_{k+1} = \mathrm{L}\}$ of $[0, \mathrm{L}]$ becomes finer and finer; see Section 6.2.

4.3. Azéma noise and quantum Wiener and Poisson processes. Applying Theorem 4.2.2 and Prop. 4.2.4 to the commutation factor case $\mathcal{T}(V) = \mathbb{C}\langle x, x^*\rangle$, see Section 1.6, we can describe the additive q-white noise as follows. Let there be given

- a family $(D_{(m)})_{m\in\mathbb{Z}}$ of pre-Hilbert spaces
- a family $(w_m)_{m\in\mathbb{Z}}$ of operators with $w_m \in \mathbf{H}(D_{(m)}, D_{(m+1)})$
- vectors $l \in D_{(1)}$ and $\tilde{l} \in D_{(-1)}$
- a complex number h with $h = 0$ unless $q = 1$.

We put $D = \bigoplus_{m\in\mathbb{Z}} D_{(m)}$, $w = \oplus_{m\in\mathbb{Z}} w_m$ and $\alpha = \oplus_{m\in\mathbb{Z}} q^m$. Then

$$F_t = \int_0^t (\mathrm{d}A_r^*(l) + \mathrm{d}\Lambda_r(w) + \mathrm{d}A_r(\tilde{l}) + h\mathrm{d}r)\Gamma_{r,\infty}(\alpha)$$

is an additive q-white noise with generator given by $(\eta_0, \rho_0, \psi_0, \alpha)$ where $\eta_0(x) = l$, $\eta_0(x^*) = \tilde{l}$, $\rho_0(x) = w$, $\psi_0(x) = h$ and $\alpha(x) = \alpha$. Conversely, any q-additive white noise is equivalent to one of this form. More precisely, if ψ is a generator we can put $D_{(m)} = \eta_\psi(\mathbb{C}\langle x, x^*\rangle_{(m)})$, $w_m = \rho_\psi(x)\lceil D_{(m)}$, $l = \eta_\psi(x)$, $\tilde{l} = \eta_\psi(x^*)$ and $h = \psi(x)$. Moreover, $F_t = B_t\Gamma_{t,\infty}(\alpha)$ with B_t the solution of

$$B_t - (A_t^*(l) + \Lambda_t(w) + A_t(\tilde{l}) + ht) = \int_0^t B_r \mathrm{d}\Lambda_r(\alpha - \mathrm{id}).$$

Here is a class of examples of additive white noise. Suppose Q is a positive sesquilinear form on double-module $*$-vector space V satisfying

$$Q(v.\lambda^*, \lambda.w) = Q(v, w) \text{ for all } \lambda \in \mathbb{L}, v, w \in V. \tag{4.3.1}$$

Furthermore, let φ be an α-invariant positive linear functional on $\mathcal{T}(V)$. We define the linear functional $g_{Q,\varphi}$ on $\mathcal{T}(V)$ by

$$g_{Q,\varphi}(1) = g_{Q,\varphi}(v) = 0$$
$$g_{Q,\varphi}(vbw) = Q(v^*, w)\varphi(b)$$

for $v, w \in V$ and $b \in \mathcal{T}(V)$; see [80].

4.3.1. PROPOSITION. *The linear functional $g_{Q,\varphi}$ is a generator with D, η_0 and ρ_0 given by*

$$D = E_Q \otimes E_\varphi$$
$$\eta_0(v) = \eta_Q(v) \otimes \Omega_\varphi$$
$$\rho_0(v) = \mathrm{id} \otimes \pi_\varphi(v)$$

where $E_Q = V/\mathcal{N}_Q$ with $\mathcal{N}_Q$ the null space of Q, η_Q is the canonical mapping from V tc E_Q and $(E_\varphi, \Omega_\varphi, \pi_\varphi)$ is the GNS-triplet associated with φ.

PROOF: It is clear that $g_{Q,\varphi}$ is α-invariant and hermitian. An element $b \in K^1$ always can be written in the form

$$b = \sum_{k=1}^m b_k v_k$$

with $m \in \mathbb{N}$, $v_1, \ldots, v_m \in V$ and $b_1, \ldots, b_m \in \mathcal{T}(V)$. We have

$$\begin{aligned} g_{Q,\varphi}(b^*b) &= \sum_{k,l=1}^{m} g_{Q,\varphi}(v_k^* b_k^* b_l v_l) \\ &= \sum_{k,l=1}^{m} Q(v_k, v_l)\varphi(b_k^* b_l) \\ &\geq 0 \end{aligned}$$

because the Schur product of two positive definite matrices is again positive definite. This proves that $g_{Q,\varphi}$ is a generator. Finally, we have for $z_1, z_2 \in \mathbb{C}$, $b_1, b_2 \in \mathcal{T}(V)$ and $v_1, v_2 \in V$

$$\begin{aligned} \langle \eta(z_1\mathbf{1} + b_1 v_1), \eta(z_2\mathbf{1} + b_2 v_2)\rangle &= g_{Q,\varphi}((b_1 v_1)^* b_2 v_2) \\ &= Q(v_1, v_2)\varphi(b_1^* b_2) \\ &= \langle \eta_Q(v_1), \eta_Q(v_2)\rangle\langle \pi_\varphi(b_1)\Omega_\varphi, \pi_\varphi(b_2)\Omega_\varphi\rangle \end{aligned}$$

from which we deduce the indicated form of D, η_0 and ρ_0.$\diamondsuit$

We investigate the structure of an additive white noise with generator $g_{Q,\varphi}$ in the special case when φ is a $*$-algebra homomorphism. Then φ is determined by a hermitian linear functional β on V. i.e. $\varphi = \mathcal{T}(\beta)$. Moreover, $E_\varphi = \mathbb{C}$ and $\pi_\varphi = \varphi$, so that $\eta(v) = \eta_Q(v)$ and $\rho(v) = \beta(v)$, and we have

$$F_t(v) = \int_0^t (\mathrm{d}A_\tau^*(\eta_Q(v)) + \beta(v)\mathrm{d}\Lambda(\mathrm{id}) + \mathrm{d}A_\tau(\eta_Q(v^*)))\Gamma_{\tau,\infty}(\alpha_{\mathrm{d}(v)}).$$

Assume for a moment that α and γ are trivial, and let $(v_i)_{i\in I}$ be a vector space basis of V consisting of hermitian elements. Then

$$F_t(v_i) = W_{t,i} + \beta_i N_t$$

with

$$\begin{aligned} W_{t,i} &= A_t^*(\eta_Q(v_i)) + A_t(\eta_Q(v_i)), \\ \beta_i &= \beta(v_i),\ N_t = \Lambda_t(\mathrm{id}). \end{aligned}$$

If $\#I = n < \infty$, the generator on $\mathbb{C}[v_1, \ldots, v_n] \subset \mathbb{C}\langle v_1, \ldots, v_n\rangle$ of the corresponding process on $\mathbb{C}[v_1, \ldots, v_n]$ of an n-dimensional Wiener process with diffusion matrix Q and drift vector 0 is given by

$$\begin{aligned} g_Q(v_1^{l_1} \ldots v_m^{l_n}) &= 0 \text{ for } \sum_{k=1}^{m} l_k \neq 2 \\ g_Q(v_k v_l) &= Q_{kl}. \end{aligned}$$

Thus $g_Q = g_{Q,\varepsilon}$. A non-commutative generalization is obtained if we drop the property of the classical covariance matrix Q to be symmetric and allow Q to be an arbitrary positive complex sesquilinear form. The process given by g_Q is $F_t(v_i) = W_{t,i}$, $i \in I$.

Each $W_{t,i}$ can be identified with a 1-dimensional Wiener process via the Segal duality transformation; see [36]. However, the $W_{t,i}$ do not commute for different i if Q is not symmetric.

The generator on $\mathbb{C}[x]$ of a Poisson process with intensity $\kappa > 0$ and with jumps of length $\beta \in \mathbb{R}$ is equal to $\kappa(\mathcal{T}(\beta) - \varepsilon)$. This is also of the form $g_{Q,\varphi}$. It is the special case when $V = \mathbb{C}$, the positive linear functional φ is $\mathcal{T}(\beta)$ and $Q = \kappa\beta^2$. Let $\#I = n < \infty$. We call the process given by $g_{Q,\mathcal{T}(\beta)}$ a *non-commutative Poisson procees* if $Q \neq 0$ and $\beta \neq 0$. This is motivated by the following considerations. We can assume that $\beta(v_i) \neq 0$, if not so, we can pass to some other basis of V. We can write Q as the sum of positive definite matrices of rank 1. Thus $g_{Q,\mathcal{T}(\beta)}$ is the sum of generators $g_{\kappa,\beta}$ with $\kappa = (\kappa_1, \ldots, \kappa_n) \in \mathbb{C}^n$ and $\beta = (\beta_1, \ldots, \beta_n) \in \mathbb{R}^n$, $\beta_k \neq 0$, and

$$g_{\kappa,\beta}(v_{i_1} \ldots v_{i_m}) = \begin{cases} 0 & \text{if } m = 1 \\ \overline{\kappa_{i_1}}\beta_{i_2} \ldots \beta_{i_{m-1}}\kappa_{i_m} & \text{if } m \geq 2. \end{cases}$$

The process given by $g_{\kappa,\beta}$ is

$$F_t(v_i) = \kappa_i A_t^* + \bar{\kappa}_i A_t + \beta_i N_t.$$

For each $i \in \{1, \ldots, n\}$ the restriction of this process to $\mathrm{Lin}\{v_i\}$ is a centralized Poisson process of intensity $|\kappa_i|^2/\beta_i^2$ and with jumps of length κ_i; cf. [41]. However, in general the realizations $F_t(v_i)$ of the Poisson processes on Fock space do not commute for different i.

We return to $V = \mathrm{Lin}\{x, x^*\}$ and the commutation factor case with $q \in \mathbb{C}_0$. If $q = 1$ put $v_1 = \frac{1}{\sqrt{2}}(x^* + x)$, $v_2 = \frac{-i}{\sqrt{2}}(x^* - x)$. Then for $Q = \frac{1}{2}\begin{pmatrix} 1 & -i \\ i & 1 \end{pmatrix}$ we have $W_{t,1} = \frac{1}{\sqrt{2}}(A_t^* + A_t)$ and $W_{t,2} = \frac{-i}{\sqrt{2}}(A_t^* - A_t)$. This is the quantum Wiener process of A.M. Cockcroft and R.L. Hudson [22] in the Bose case. In the general q-case the process given by g_Q is

$$F_t(x) = \int_0^t \mathrm{d}A_\tau \Gamma_{\tau,\infty}(q)$$

and $F_t(x) = L_t(x)\Gamma_{t,\infty}(q)$ with $L_t(x)$ the solution of

$$L_t(x) - A_t = (q-1)\int_0^t L_\tau(x)\mathrm{d}\Lambda_\tau. \tag{4.3.2}$$

The corresponding equations for the left q-white noise are

$$\check{L}_t(x) = \check{F}_t(x) = \int_0^t \Gamma_\tau(q)\mathrm{d}A_\tau$$

and

$$\check{L}_t(x) - A_t = \check{F}_t(x) - A_t = (q-1)\int_0^t \mathrm{d}\Lambda_\tau \check{F}_\tau(x).$$

According to the remarks following Theorem 4.2.2, we also have

$$L_t(x) = \bar{q}\Big(\int_0^t \Gamma_\tau(q^{-1})\mathrm{d}A_\tau\Big)\Gamma_t(q) = \Gamma_t(q)\Big(\int_0^t \Gamma_\tau(q^{-1})\mathrm{d}A_\tau\Big) \tag{4.3.3}$$

where we used that η, ρ and ψ vanish on x. In particular, for $q = -1$ we have

$$\check{F}_t(x) = \int_0^t \Gamma_r(-1)\mathrm{d}A_r$$

$$F_t(x) = \Gamma_{0\infty}(-1)\Big(\int_0^t \Gamma_r(-1)\mathrm{d}A_r\Big) = \Gamma_{0\infty}(-1)\check{F}_t(x)$$

which again shows that F_t and $\check{F}_t$ are equivalent in the case $q = -1$.

Equations (4.3.2) and (4.3.3) appear in the quantum probabilistic treatment of the Azéma martingale; see [61]. For $q \in \mathbb{R}$ with $|q| \leq 1$ the process $X_t = L_t^* + L_t$ with $L_t = L_t(x)$ gives rise to a commuting family of self-adjoint operators which, in the vacuum state, is the Azéma martingale with parameter q. Classically, X_t appear as square integrable martingales with sample paths which are right continuous and have left limits such that

$$\mathrm{d}[X,X]_t = (q-1)X_{t-}\mathrm{d}X_t + \mathrm{d}t$$

where $[X,X]_t$ denotes the quadratic variation of X_t. For the 'degenerate' case $q = 0$ J. Azéma proved the existence of such a process in [10]. Then M. Eméry [26] constructed X_t for arbitrary real q and proved that, for $|q| \leq 1$, the martingale X_t has the chaos completeness property. The processes X_t were the first examples of martingales with this property which are not classical stochastic processes with independent increments. However, we have seen that X_t can be written as a sum $L_t^* + L_t$ where the pair (L_t^*, L_t) constitutes a *quantum* white noise (on $\mathcal{T}(V)_{(\gamma,\alpha)}$) that is a quantum stochastic process with independent increments. Notice that it also follows from Prop. 4.2.3 that X_t is a commutative process. However, $\check{X}_t = \check{L}_t^* + \check{L}_t$ is not commutative in the case when $q \neq \pm 1$. For example, for $t_1 < t_2$, $q \in \mathbb{R}$

$$\begin{aligned}\langle\Omega, \check{X}_{t_1}(\check{X}_{t_2})^2\check{X}_{t_1}\Omega\rangle - \langle\Omega, (\check{X}_{t_1})^4\Omega\rangle &= \langle\Omega, \check{X}_{t_1}(\Gamma_{t_1}(q)\check{X}_{t_1t_2})^2\check{X}_{t_1}\Omega\rangle \\ &= \langle\Omega, \check{L}_{t_1}(\Gamma_{t_1}(q)\check{X}_{t_1t_2})^2\check{L}_{t_1}^*\Omega\rangle \\ &= q^2\langle\Omega, \check{L}_{t_1}\check{L}_{t_1}^*\Omega\rangle\langle\Omega, (\check{X}_{t_1t_2})^2\Omega\rangle \\ &= q^2 t_1(t_2 - t_1)\end{aligned}$$

whereas

$$\begin{aligned}\langle\Omega, (\check{X}_{t_2})^2(\check{X}_{t_1})^2\Omega\rangle - \langle\Omega, (\check{X}_{t_1})^4\Omega\rangle &= \langle\Omega, (\Gamma_{t_1}(q)\check{X}_{t_1t_2})^2\check{L}_{t_1}\check{L}_{t_1}^*\Omega\rangle \\ &= t_1(t_2 - t_1).\end{aligned}$$

Furthermore, F_t and $\check{F}_t$ are not equivalent. For example (cf. Prop. 1.9.3),

$$\begin{aligned}\langle\Omega, F_{t_1}F_{t_2}F_{t_2}^*\Omega\rangle - \langle\Omega, (F_{t_1})^2(F_{t_1}^*)^2\Omega\rangle &= \langle\Omega, F_{t_1}F_{t_1}^*\Omega\rangle\langle\Omega, F_{t_2}F_{t_2}^*\Omega\rangle \\ &= t_1t_2\end{aligned}$$

but

$$\begin{aligned}\langle\Omega, \check{F}_{t_1}\check{F}_{t_2}\check{F}_{t_2}^*\check{F}_{t_1}^*\Omega\rangle - \langle\Omega, (\check{F}_{t_1})^2(\check{F}_{t_2}^*)^2\Omega\rangle &= \langle\Omega, \check{F}_{t_1}\check{F}_{t_1t_2}\check{F}_{t_1t_2}^*\check{F}_{t_1}^*\Omega\rangle \\ &= q^{-2}t_1t_2.\end{aligned}$$

We look at a general covariance matrix Q which we now write in the form

$$Q = \begin{pmatrix} g_Q(xx^*) & g_Q(x^2) \\ g_Q((x^2)^*) & g_Q(x^*x) \end{pmatrix}$$

and see that unless $q = \pm 1$ the invariance condition (4.3.1) forces the off-diagonal entries to be 0. If $q = \pm 1$ there is no further restriction on Q than that of positive definiteness. We treated the case $q = 1$, for the case $q = -1$ see Section 4.5 on cocommutative white noise. Let $q \neq \pm 1$. Then $Q = \begin{pmatrix} \kappa_1 & 0 \\ 0 & \kappa_2 \end{pmatrix}$ with $\kappa_i \geq 0$, $i = 1, 2$. The equations for g_Q are

$$F_t(x) = \int_0^t (\mathrm{d}A_\tau^* \begin{pmatrix} 0 \\ \sqrt{\kappa_2} \end{pmatrix} + \mathrm{d}A_\tau \begin{pmatrix} \sqrt{\kappa_1} \\ 0 \end{pmatrix})\Gamma_{\tau,\infty} \begin{pmatrix} q & 0 \\ 0 & q^{-1} \end{pmatrix}$$

and

$$L_t(x) - (A_t^* \begin{pmatrix} 0 \\ \sqrt{\kappa_2} \end{pmatrix} + A_t \begin{pmatrix} \sqrt{\kappa_1} \\ 0 \end{pmatrix}) = \int_0^t L_\tau(x) \mathrm{d}\Lambda_\tau \begin{pmatrix} q-1 & 0 \\ 0 & q^{-1}-1 \end{pmatrix}.$$

We call (L_t^*, L_t) 2-dimensional *Azéma noise* with variance (κ_1, κ_2). Since we have that $(\check{L}_t^*, \check{L}_t)$ for $q = 1$ is a 2-dimensional Bose quantum Wiener process and for $q = -1$ is a 2-dimensional Fermi Wiener process (see Section 4.5 and [74]), we call $(\check{L}_t^*, \check{L}_t)$ a *q-interpolation between Bose and Fermi quantum Brownian motion.*

4.4. Multiplicative and unitary white noise. We begin this section with some introductory remarks. We show that for a symmetric $*$-bialgebra $\mathcal{B}$ each element in $\mathcal{B}$ lies in a sub-$*$-bialgebra of $\mathcal{B}$ which is the quotient of a $*$-bialgebra of type $\mathcal{M}\langle n \rangle$ by an appropriate $*$-bi-ideal.

To prove this we develop some aspects of the theory of the $*$-bialgebra $\mathcal{R}(G)$ formed by the representative functions on a semi-group G; cf. Section 1.7. Let $\mathcal{R}(G)$ denote the vector space formed by the complex-valued functions f on G with the property that there exists an $n \in \mathbb{N}$ and functions $f_1, \ldots, f_n, g_1, \ldots, g_n \in \mathbb{C}^G$ such that

$$f(xy) = \sum_{k=1}^n f_k(x) g_k(y) \text{ for all } x, y \in G;$$

see [1]. For $x, y \in G$ and $f \in \mathbb{C}^G$ denote by f^x, f_y and f_y^x the translates of f given by $(f^x)(z) = f(zx)$, $(f_y)(z) = f(yz)$ and $f_y^x = (f_y)^x = (f^x)_y$ respectively. Clearly, $f \in \mathcal{R}(G)$ if and only if $f_y^x \in \mathcal{R}(G)$ for all $x, y \in G$. We put $\mathcal{R}_f(G) = \mathrm{Lin}\{f_y^x : x, y \in G\}$ and $\mathcal{R}_f^{\mathrm{r}}(G) = \mathrm{Lin}\{f_y : y \in G\}$. Then

4.4.1. LEMMA. *A complex-valued function f on G is in $\mathcal{R}(G)$ if and only if there exist $n \in \mathbb{N}$, $x_1, \ldots, x_n \in G$ and $g_1, \ldots, g_n \in \mathcal{R}_f^{\mathrm{r}}(G)$ such that*

$$f(xy) = \sum_{k=1}^n (f^{x_k})(x) g_k(y) \text{ for all } x, y \in G. \tag{4.4.1}$$

PROOF: If (4.4.1) holds then $f \in \mathcal{R}(G)$.

Suppose $f \in \mathcal{R}(G)$ and $f(xy) = \sum f_k(x)g_k(y)$. We can assume without loss of generality that $f_1, \ldots, f_n$ are linearly independent and that there exist $y_1, \ldots, y_n \in G$ such that $f_k(y_l) = \delta_{kl}$. Then

$$f(y_l x) = \sum_{k=1}^{n} f_k(y_l)g_k(x) = g_l(x)$$

and $g_l = f_{y_l}$, so that we have $f(xy) = \sum f_k(x)g_k(y)$ with $g_k \in \mathcal{R}^{\mathrm{r}}_f(G)$. Next we can assume without loss of generality that $g_1, \ldots, g_n \in \mathcal{R}^{\mathrm{r}}_f(G)$ are linearly independent and that there exist $x_1, \ldots, x_n \in G$ such that $g_k(x_l) = \delta_{kl}$. Then

$$f(xx_l) = \sum_{k=1}^{n} f_k(x)g_k(x_l) = f_l(x)$$

and we have

$$f(xy) = \sum_{k=1}^{n} (f^{x_k})(x)g_k(y)$$

with $g_k \in \mathcal{R}^{\mathrm{r}}_f(G)$ as desired.◇

Let ι denote the embedding of $\mathbb{C}^G \otimes \mathbb{C}^G$ into $\mathbb{C}^{G\times G}$ given by $\iota(f \otimes g)(x, y) = f(x)g(y)$. Lemma 4.4.1 yields

$$\Delta\mathcal{R}(G) \subset \iota(\mathcal{R}(G) \otimes \mathcal{R}(G))$$

where $\Delta : \mathbb{C}^G \to \mathbb{C}^{G\times G}$ is given by $(\Delta f)(x, y) = f(xy)$. We identify $\mathcal{R}(G) \otimes \mathcal{R}(G)$ with its image in $\mathbb{C}^{G\times G}$ which actually is equal to $\mathcal{R}(G \times G)$. The $*$-algebra $\mathcal{R}(G)$ of functions on G becomes a symmetric $*$-bialgebra with comultiplication Δ and with counit given by $\epsilon(f) = f(\mathrm{e})$. Moreover, $\mathcal{R}_f(G)$ is a subcoalgebra of $\mathcal{R}(G)$ and it is the smallest subcoalgebra of $\mathcal{R}(G)$ containing f.

The importance of coalgebras of type $\mathcal{R}(G)$ lies in the following construction; see [1]. Let $\mathcal{C}$ be a coalgebra. Then $\mathcal{C}^*$ is a semi-group with multiplication given by the convolution product of linear functionals. One checks that $(\mathcal{C}^*)^* \cap \mathcal{R}(\mathcal{C}^*)$ is a subcoalgebra of $\mathcal{R}(\mathcal{C}^*)$. We have the canonical embedding of $\mathcal{C}$ into $(\mathcal{C}^*)^*$ and $\mathcal{C} \subset (\mathcal{C}^*)^* \cap \mathcal{R}(\mathcal{C}^*)$. The coalgebra $\mathcal{C}$ can be regarded as a subcoalgebra of $(\mathcal{C}^*)^* \cap \mathcal{R}(\mathcal{C}^*)$. Since the intersection of coalgebras is again a coalgebra, we obtain for $c \in \mathcal{C}$ that $\mathcal{R}_c(\mathcal{C}^*) \subset \mathcal{C}$, and $\mathcal{R}_c(\mathcal{C}^*)$ is the smallest subcoalgebra of $\mathcal{C}$ containing c.

4.4.2. THEOREM. *Let $\mathcal{C}$ be a coalgebra and let $c \in \mathcal{C}$. Then there exist $n \in \mathbb{N}$ and $c_{kl} \in \mathcal{C}$, $k, l = 1, \ldots, n$, such that*

$$\Delta c_{kl} = \sum_{m=1}^{n} c_{km} \otimes c_{ml},$$

$$\epsilon c_{kl} = \delta_{kl}$$

and

$$\mathcal{R}_c(\mathcal{C}^*) = \mathrm{Lin}\{c_{kl} : k, l =, 1, \ldots, n\}.$$

PROOF: By our above remarks it is sufficient to treat the case $C = \mathcal{R}_f(G)$, $c = f$. By Lemma 4.4.1

$$\Delta f = \sum_{k=1}^{n} f^{x_k} \otimes g_k$$

with $g_k \in \mathcal{R}_f^{\mathrm{r}}(G)$. By dropping some of the f^{x_k} and changing the g_k if necessary, we can assume that $f^{x_1}, \ldots, f^{x_n}$ are linearly independent. Define the representation π of G on the vector space $\mathcal{R}_f(G)$ by $\pi(x)g = g^x$. Then

$$\pi(x)(f^{x_l}) = f^{xx_l} = \sum_{k=1}^{n} f^{x_k} g_k(xx_l)$$

which shows that π leaves invariant the linear subspace V of $\mathcal{R}_f(G)$ spanned by the f^{x_k}, $k = 1, \ldots, n$. The matrix of $\pi(x)$ restricted to V with respect to the basis $\{f^{x_1}, \ldots, f^{x_n}\}$ is given by $\pi(x)_{kl} = g_k(xx_l)$. We have that the mappings $x \mapsto \pi(x)_{kl}$ are in $\mathcal{R}_f(G)$, and

$$\pi(xy)_{kl} = \sum_{m=1}^{n} \pi(x)_{km}\pi(y)_{ml},$$

$$\pi(\mathrm{e})_{kl} = \delta_{kl}.$$

Moreover,

$$\begin{aligned}\sum_{k,l} f_k(\mathrm{e})g_l(\mathrm{e})\pi(x)_{kl} &= \sum_{k,l} f_k(\mathrm{e})g_k(xx_l)g_l(\mathrm{e})\\ &= \sum_{l}(f^{x_l})(x)g_l(\mathrm{e})\\ &= f(x),\end{aligned}$$

so that $c_{kl} = \pi(\cdot)_{kl}$ have the desired properties.◇

Notice that we proved the fundamental theorem on coalgebras.

Let $\mathcal{B}$ be a symmetric $*$-bialgebra and let $b \in \mathcal{B}$. By Theorem 4.4.2 we can find $n \in \mathbb{N}$ and $b_{kl} \in \mathcal{B}$ such that $\Delta b_{kl} = \sum b_{km} \otimes b_{ml}$, $\epsilon b_{kl} = \delta_{kl}$ and $b \in \mathrm{Lin}\{b_{kl} : k, l = 1, \ldots, n\}$. Then $\Theta x_{kl} = b_{kl}$ defines a homomorphism between $\mathcal{M}\langle n\rangle$ and the sub-$*$-bialgebra of $\mathcal{B}$ generated by the b_{kl}. Thus our introductory claim follows. As a consequence, a white noise on an arbitrary symmetric $*$-bialgebra 'locally' gives rise to a white noise on $\mathcal{M}\langle n\rangle$ if restricted to the sub-$*$-bialgebra generated by the b_{kl}. However, in passing to $\mathcal{M}\langle n\rangle$, the specific structure of our white noise on $\mathcal{B}$ can no longer be seen and we loose information. An extreme example is that of tensor Hopf algebras.

We investigate white noise on $\mathcal{M}\langle n\rangle$ and begin with the symmetric case. For a generator ψ on the symmetric $*$-bialgebra $\mathcal{M}\langle n\rangle$, see Section 1.6, we define $L, \tilde{L} \in M_n \otimes D$, $W \in M_n \otimes \mathbf{H}(D)$ and $K \in M_n$ by

$$L_{kl} = \eta(x_{kl});\ \tilde{L}_{kl} = \eta(x_{lk}^*)$$

$$W_{kl} = \rho(x_{kl})$$

$$K_{kl} = \psi(x_{kl}).$$

Moreover, denote by $\mathrm{d}A_t^*(L)$, $\mathrm{d}A_t(\tilde{L})$ and $\mathrm{d}\Lambda_t(W-\mathrm{id})$ the matrices of differentials given by

$$(\mathrm{d}A_t^*(L))_{kl} = \mathrm{d}A_t^*(L_{kl}); \ (\mathrm{d}A_t(\tilde{L}))_{kl} = \mathrm{d}A_t(\tilde{L}_{lk}),$$
$$(\mathrm{d}\Lambda_t(W-\mathrm{id}))_{kl} = \mathrm{d}\Lambda_t(W_{kl} - \delta_{kl}\mathrm{id})$$

respectively. Given $(J_{st})_{kl} \in \underline{A}_D$, $k,l = 1,\ldots,n$, and $L \in M_n \otimes D$ we write the n^2 equations

$$\int \sum_{m=1}^{n} (J_{sr})_{km} \mathrm{d}A_r^*(L)_{ml}$$

in the more compact form

$$\int J_{sr} \mathrm{d}A^*(L).$$

Similarly for the other types of integrals. An application of Theorem 2.5.3 yields the following result.

4.4.3. THEOREM. *Let ψ be a generator on $\mathcal{M}\langle n\rangle$. Then the equation*

$$J_{st} = \mathrm{id} + \int_s^t J_{sr}(\mathrm{d}A_r^*(L) + \mathrm{d}\Lambda_r(W-\mathrm{id}) + \mathrm{d}A_r(\tilde{L}) + K\mathrm{d}r) \qquad (4.4.2)$$

on $\mathbb{C}^n \otimes \mathcal{F}(L^2(\mathbb{R}_+, H))$ has a unique solution J_{st} with $(J_{st})_{kl} \in \underline{A}_D$. In the vacuum state, J_{st} is a realization of a white noise on $\mathcal{M}\langle n\rangle$ with generator ψ.$\diamondsuit$

The converse also holds. Given $D, L, \tilde{L}$, W and K, equation (4.4.2) has a unique solution and this solution is a multiplicative white noise. Its generator can be computed in analogy to Prop. 4.2.4.

In the case of unitary white noise we are dealing with bounded operators. It is natural to ask if there is a unique *unitary* solution of our quantum stochastic differential equation. On the other hand, it is not true in general that the components of a unitary operator on $\mathbb{C}^n \otimes \mathcal{H}$ are given by kernels; cf. [49]. However, the class of integrand processes in the original quantum stochastic calculus of Hudson and Parthasarathy consists of *all adapted locally square integrable* processes on $\mathcal{K} \otimes \mathcal{F}(L^2(\mathbb{R}_+, H))$ where $\mathcal{K}$ is the 'initial Hilbert space' and where a process $(J_{st})_{(s,t)\in T}$ on $\mathcal{K} \otimes \mathcal{F}(L^2(\mathbb{R}_+, H))$ is called locally square integrable if the functions

$$t \mapsto J_{st}(u \otimes \mathrm{E}(f))$$

are square integrable on all bounded intervals for all s and all $f \in L^2(\mathbb{R}_+) \otimes H$, $u \in \mathcal{K}$. Notice that our space of 'admissible functions' in the sense of [41] is $L^2(\mathbb{R}_+) \otimes H$. Clearly, any process consisting of unitary operators is locally square integrable. For our purposes $\mathcal{K} = \mathbb{C}^n$. We denote by $\mathcal{L}^2(\mathbb{C}^n, H)$ the space of all adapted locally square integrable processes. We have that $(J_{st}) \in \mathcal{L}^2(\mathbb{C}^n, H)$ if and only if it is adapted and if $((J_{st})_{kl}) \in \mathcal{L}^2(\mathbb{C}, H)$ for all $k,l = 1,\ldots,n$. The following shows that $(J_{st}) \in \mathcal{L}^2(\mathbb{C}^n, H)$ if $((J_{st})_{kl}) \in \underline{A}_D$ for all $k,l = 1,\ldots,n$.

4.4.4. PROPOSITION. *For* $(j_{st}) \in \underline{A}_D$ *the functions* $t \mapsto j_{st}F$ *are square integrable on bounded intervals for all* $F \in \mathcal{E}_D$.

PROOF: By Lemma 3.2.1

$$\int_s^L \|j_{st}F\|^2 dt \leq \int_s^L (3^\omega (\sum_{\sigma \dot\cup \tau \dot\cup \varepsilon = \omega} \int_S \|j_{st}(\sigma, \tau, \varrho) F(\varepsilon \cup \tau \cup \varrho)\| d\varrho)^2 d\omega) dt$$
$$\leq (L - s) e^{(L-s)27c^2} \int_S (27c^2)^\omega \|F(\omega)\|^2 d\omega$$

where c is the common constant of (K3) for j_{st} in the interval $[s, L]$. This completes the proof of the proposition.◇

The domains of the Hudson-Parthasarathy integrals consist of spans of sets of exponential vectors. The following shows that the kernel integrals and the Hudson-Parthasarathy integrals agree on the intersection of their domains.

4.4.5. PROPOSITION. *Let* $x_{st}^{(i)}$ *be in* $\mathcal{A}_D$, $i = 1, 2, 3, 4$, *and* $\xi, \tilde{\xi} \in D$, $R \in \mathbf{H}(D)$. *Then for*

$$k_{st} = \int_s^t (x_{sr}^{(1)} da_r^*(\xi) + x_{sr}^{(2)} d\lambda_r(R) + x_{sr}^{(3)} da_r(\tilde{\xi}) + x_{sr}^{(4)} dr)$$

the equation

$$\begin{aligned} &\langle E(f), \underline{k}_{st} E(g)\rangle \\ &= \int_s^t (\langle E(f), \underline{x}_{sr}^{(1)} E(g)\rangle \langle f(r), \xi\rangle + \langle E(f), \underline{x}_{sr}^{(2)} E(g)\rangle \langle f(r), Rg(r)\rangle \\ &\quad + \langle E(f), \underline{x}_{sr}^{(3)} E(g)\rangle \langle \tilde{\xi}, g(r)\rangle + \langle E(f), \underline{x}_{sr}^{(4)} E(g)\rangle) dr \end{aligned} \tag{4.4.3}$$

holds for all functions $f, g \in L^2(\mathbb{R}_+) \otimes D$.

PROOF: It suffices to show (4.4.3) in the cases when only one of the integrand processes $\underline{x}_{st}^{(i)}$ is not equal to 0. For example, if $\underline{x}_{st}^{(i)} = 0$ for $i = 2, 3, 4$, we have

$$\begin{aligned} &\langle E(f), \underline{k}_{st} E(g)\rangle \\ &= \int_S \langle \otimes_{u\in\omega}^{\rightarrow} f(u), \sum_{\sigma \dot\cup \tau \dot\cup \varepsilon = \omega} (\int_S k_{st}(\sigma, \tau, \varrho) \otimes_{u \in \tau \dot\cup \varepsilon \dot\cup \varrho}^{\rightarrow} g(u) d\varrho)\rangle d\omega \\ &= \int_s^t (\int_S \langle \otimes_{u\in\omega}^{\rightarrow} f(u), \sum_{\sigma \dot\cup \tau \dot\cup \varepsilon = \omega} (\int_S x_{sr}^{(1)}(\sigma, \tau, \varrho) \otimes_{u \in \tau \dot\cup \varepsilon \dot\cup \varrho}^{\rightarrow} g(u) d\varrho\rangle d\omega) \langle f(r), \xi\rangle dr \\ &= \int_s^t \langle E(f), \underline{x}_{sr}^{(1)} E(g)\rangle \langle f(r), \xi\rangle dr. \end{aligned}$$

Similarly for the remaining cases.◇

For $L \in M_n \otimes D$ we denote by $\langle L, L\rangle$ the element in M_n with

$$(\langle L, L\rangle)_{kl} = \sum_{m=1}^n \langle L_{mk}, L_{ml}\rangle.$$

4.4.6. THEOREM. *Let H be a Hilbert space, $n \in \mathbb{N}_0$, and let $L \in M_n \otimes H$, $W \in M_n \otimes \mathbf{B}(H)$ and $R \in M_n$ such that W is unitary and R is self-adjoint. Then the Hudson-Parthasarathy quantum stochastic integral equation*

$$U_t = \mathrm{id} + \int_0^t U_\tau \mathrm{d}I_\tau \tag{4.4.4}$$

with

$$I_t = A_t^*(L) + \Lambda_t(W - \mathrm{id}) - A_t(W^* L) + (\mathrm{i}R - \frac{1}{2}\langle L, L\rangle)t)$$

has a unique solution in $\mathcal{L}^2(\mathbb{C}^n, H)$. Moreover, the U_t, $t \geq 0$, extend to unitary operators on $\mathbb{C}^n \otimes \mathcal{F}(L^2(\mathbb{R}_+, H))$.

PROOF: We have that $U_N(t)$, $N \in \mathbb{N}$, $t \geq 0$, given by the Picard-Lindelöf iteration procedure

$$U_0(t) = \mathrm{id}$$

$$U_{N+1}(t) = \mathrm{id} + \int_0^t U_N(s)\mathrm{d}I_s$$

satisfy for $\mathrm{L} \in \mathbb{R}_+$ and $f \in L^2(\mathbb{R}_+) \otimes H$ locally bounded

$$\sup_{\substack{0 \leq t \leq \mathrm{L} \\ 1 \leq m \leq n}} \|(U_N(t) - U_{N-1}(t))e_m \otimes \mathrm{E}(f)\| \leq \mathrm{e}^{\frac{1}{2}(\mathrm{L}+\|f\|^2)} \frac{2n\sqrt{6}\mathrm{L}c(\mathrm{L}))^N}{\sqrt{N!}}. \tag{4.4.5}$$

Here $\{e_1, \ldots, e_n\}$ is a basis of $\mathbb{C}^n$ and $c(\mathrm{L})$ denotes the supremum of the real numbers $|\langle f(t), W_{kl} - \delta\mathrm{id}) f(t)\rangle|$, $|\langle f(t), L_{kl}\rangle|$, $|\langle f(t), (W^* L)_{kl}\rangle|$, $\|(W_{kl} - \delta_{kl}\mathrm{id}) f(t)\|^2$, $\|(W^* L)_{kl}\|^2$ and $|\mathrm{i}R_{kl} - \frac{1}{2}\langle L, L\rangle_{kl}|$ where t varies through the interval $[0, \mathrm{L}]$ and k and l run from 1 to n. This follows from

$$\begin{aligned}
&\|(U_N(t) - U_{N-1}(t))e_m \otimes \mathrm{E}(f)\| \\
&= \|\sum_{k,l=1}^n \int_0^t (\sum_{u=1}^n e_{ul} \otimes (U_{N-1}(s) - U_{N-2}(s))_{uk})(\mathrm{d}I_s)_{kl}) e_m \otimes \mathrm{E}(f)\| \\
&\leq 2\sqrt{6}c(\mathrm{L}) \sum_{k,l=1}^n (\int_0^{\mathrm{L}} \mathrm{e}^{t-s} \|(\sum_{u=1}^n e_{ul} \otimes (U_{N-1}(s) - U_{N-2}(s))_{uk}) e_m \otimes \mathrm{E}(f)\|^2 \mathrm{d}s)^{\frac{1}{2}} \\
&= 2\sqrt{6}c(\mathrm{L}) \sum_{k=1}^n (\int_0^{\mathrm{L}} \mathrm{e}^{t-s} \|(U_{N-1}(s) - U_{N-2}(s)) e_k \otimes \mathrm{E}(f)\|^2 \mathrm{d}s)^{\frac{1}{2}} \\
&\leq n\sqrt{6}c(\mathrm{L}) (\int_0^{\mathrm{L}} \mathrm{e}^{t-s} \max_{1 \leq k \leq n} \|(U_{N-1}(s) - U_{N-2}(s)) e_k \otimes \mathrm{E}(f)\|^2 \mathrm{d}s)^{\frac{1}{2}}
\end{aligned}$$

where we made use of the estimate (4.9) of [41]. (Here (e_{kl}) denotes the system of matrix units in the basis $\{e_k\}$ of $\mathbb{C}^n$.) Based on the estimate (4.4.5), it can be shown exactly as in the proof of Prop. 7.1 of [41], that $U_N(t)$ converge to a solution U_t of (4.4.4) for $N \to \infty$. The unitarity of the U_t is shown, as in the proof of Theorem 7.1 of

[41], by looking at matrix elements and using an argument for classical linear differential equations.◇

4.4.7. PROPOSITION. *Let H, n, L, W and R be as in Theorem 4.4.6 and assume that $\{L_{kl} : k,l = 1,\ldots,n\}$ is cyclic for $\{W_{kl} : k,l = 1,\ldots,n\}$. Then the equations*

$$\rho(x_{kl}) = W_{kl} \tag{4.4.6}$$

$$\eta(x_{kl}) = L_{kl};\ \eta(x^*_{kl}) = -(W^*L)_{lk} \tag{4.4.7}$$

$$\psi(x_{kl}) = \mathrm{i}R_{kl} - \frac{1}{2}\langle L, L\rangle_{kl} \tag{4.4.8}$$

define a $*$-representation ρ of $\mathcal{U}\langle n\rangle$ on H, a 1-cocycle η with respect to ρ and a generator ψ on $\mathcal{U}\langle n\rangle$ such that $(\rho,\eta) \sim (\rho_\psi, \eta_\psi)$. Conversely, each generator ψ on $\mathcal{U}\langle n\rangle$ appears in this way. More precisely, one puts $L_{kl} = \eta_\psi(x_{kl})$, $W_{kl} = \rho_\psi(x_{kl})$ and $R_{kl} = -\mathrm{i}(\psi(x_{kl} + \frac{1}{2}\langle L, L\rangle)_{kl})$.

PROOF: Proceeding step by step the equations (4.4.6)-(4.4.8) first determine a $*$-representation $\tilde\rho$, then a 1-cocycle $\tilde\eta$ with respect to $\tilde\rho$ and finally a generator $\tilde\psi$ on $\mathcal{M}\langle n\rangle$. It remains to show that $\tilde\rho, \tilde\eta$ and $\tilde\psi$ vanish on the relations in $\mathcal{U}\langle n\rangle$. Clearly, $\tilde\rho$ does because W is unitary. Next we have

$$\begin{aligned}\sum_{m=1}^{n} \tilde\eta(x_{km}x^*_{lm}) &= \sum_{m=1}^{n}(\rho(x_{km})\tilde\eta(x^*_{lm}) + \tilde\eta(x_{km})\delta_{lm}) \\ &= -\sum_{m,u=1}^{n} W_{km}W^*_{um}L_{ul} + L_{kl} \\ &= 0\end{aligned}$$

again by the unitarity of W. Similarly, for the other type of relations. It follows that $\tilde\eta$ respects the relations. Finally,

$$\begin{aligned}\sum_{m=1}^{n} \psi(x_{km}x^*_{lm}) &= (\sum_{m=1}^{n}\langle\eta(x^*_{km}), \eta(x^*_{lm})\rangle) + \psi(x_{kl}) + \psi(x^*_{lk}) \\ &= (\sum_{m,u,v=1}^{n} \langle W^*_{um}L_{uk}, W^*_{vm}L_{vl}\rangle) - \frac{1}{2}\langle L, L\rangle_{kl} - \frac{1}{2}\overline{\langle L, L\rangle_{lk}} \\ &= 0\end{aligned}$$

and, in a similar manner, one proves that $\sum_{m=1}^{n}\psi(x_{mk}x^*_{ml}) = 0$.

If ψ is a generator, then L, W and R as defined in the second assertion of the proposition satisfy the conditions of Theorem 4.4.6, and ψ equals the generator constructed by the procedure of the first part of the proposition.◇

4.4.8. COROLLARY. *Let H, n, L, W and R be as in Theorem 4.4.6. Then, in the vacuum state, the solution of (4.4.4) is a unitary n-dimensional white noise with generator ψ given by the first part of Prop. 4.4.7. Conversely, every n-dimensional unitary white noise is equivalent to one of this form. More precisely, if ψ is the generator we take H, L, W and R of the second part of Prop. 4.4.7.*

PROOF: Let H, L, W and R be given. We can assume without loss of generality that the entries of L are cyclic for the entries of W. The generator ψ on $\mathcal{U}\langle n\rangle$ associated with H, L, W, R by Prop. 4.4.7 gives rise to a realization U_t of a unitary white noise on $\mathbb{C}_n \otimes \mathcal{F}(L^2(\mathbb{R}_+, H))$ by Theorem 2.5.3. Applying Prop. 4.4.4, 4.4.5 and the formula (4.4) in [41] for the matrix elements, this unitary white noise satisfies the Hudson-Parthasarathy equation (4.4.3). It therefore must be equal to the solution of (4.4.4) appearing in Theorem 4.4.6.◇

Notice that Corollary 4.4.8 gives also a description of classical white noise on the group U_n of unitary $n \times n$-matrices, i.e. of classical stochastic processes with values in U_n and with independent, stationary increments. Classical white noise on U_n can be viewed as a special case of n-dimensional unitary white noise. It comes from those generators on $\mathcal{U}\langle n\rangle$ which vanish on the $*$-ideal in $\mathcal{U}\langle n\rangle$ generated by the commutators

$$[x_{kl}, x_{k'l'}],\ [x^*_{kl}, x_{k'l'}],\ k, l = 1, \ldots, n,$$

and thus give rise to generators on the coefficient algebra $\mathcal{U}[n]$ of U_n. In particular, it follows that classical white noise on U_n can be realized on Bose Fock space as the solution of a quantum stochastic differential equation of the form (4.4.4). Moreover, in the classical case equation (4.4.4) can be rewritten and becomes a classical stochastic differential equation against multi-dimensional Wiener and Poisson processes; see [37].

Next we solve the problem of classifying unitary q-white noise by an application of Prop.2.1.3.

4.4.9. THEOREM. *Let ψ be a generator on $\mathcal{M}\langle n; q\rangle$. Let α denote the linear operator on $D = \bigoplus_{\mu\in\mathbb{M}} D_{(\mu)}$ which equals multiplication by q^{μ} on $D_{(\mu)}$. We define $L, \tilde{L} \in M_n \otimes D_\psi$, $W \in M_n \otimes \mathbf{H}(D_\psi)$ and $K \in M_n$ by*

$$L_{kl} = q^{k(l-k)}\eta_\psi(x_{kl}); \tilde{L}_{kl} = \eta_\psi(x^*_{lk}) \tag{4.4.9}$$

$$W_{kl} = \alpha(y^{-k})\rho_\psi(x_{kl}) \tag{4.4.10}$$

$$K_{kl} = \psi(x_{kl}).$$

Let $\tilde{\mathrm{R}}_{st}$ be the symmetric multiplicative white noise which is the solution of (4.4.2). Then, in the vacuum state, the process $(\mathrm{R}_{st})_{(s,t)\in T}$ with

$$(\mathrm{R}_{st})_{kl} = \Gamma_{s,\infty}(\alpha)^k(\tilde{\mathrm{R}}_{st})_{kl}\Gamma_{t,\infty}(\alpha)^{-l} \tag{4.4.11}$$

is a multiplicative left q-white noise with generator ψ.

PROOF: We observe that

$$L_{kl} = \eta^q_\psi(y^{-k}x_{kl});\ \tilde{L}_{kl} = \eta^q_\psi((y^{-l}x_{lk})^*)$$
$$W_{kl} = \rho^q_\psi(y^{-k}x_{kl})$$
$$K_{kl} = \psi(y^{-k}x_{kl}).$$

By Prop. 3.1.3 we have

$$j^q_{st}(y^{-k}x_{kl}) = (\tilde{\mathrm{R}}_{st})_{kl}.$$

Thus

$$\begin{aligned} j_{st}(x_{kl}) &= j^q_{st}(x_{kl})j^q_{t,\infty}(y^{k-l}) \\ &= j^q_{st}(y^k)(\tilde{\mathrm{R}}_{st})_{kl}j^q_{t,\infty}(y^{k-l}) \\ &= j^q_{s,\infty}(y^k)(\tilde{\mathrm{R}}_{st})_{kl}j^q_{t,\infty}(y^{-l}) \end{aligned}$$

which proves the theorem.◇

A similar result holds for multiplicative *right* q-white noise, one only has to replace (4.4.9) and (4.4.10) by

$$L_{kl} = \eta_\psi(x_{kl});\ \tilde{L}_{kl} = q^{k(l-k)}\eta_\psi(x^*_{lk})$$
$$W_{kl} = \alpha((y^*)^{-k})\rho_\psi(x_{kl})$$

and (4.4.11) by

$$(\mathrm{R}_{st})_{kl} = \Gamma_s(\alpha)^k(\tilde{\mathrm{R}}_{st})_{kl}\Gamma_t(\alpha)^{-l}.$$

We apply this to unitary q-white noise to obtain the result that all unitary q-white noise are of the form $U_tM_t^{-1}$ with U_t a symmetric unitary white noise and with M_t the unitary process on $\mathbb{C}^n \otimes \mathcal{F}(L^2(\mathbb{R}_+, H))$ given by

$$(M_t)_{kl} = \delta_{kl}\Gamma_t(\alpha)^k.$$

We can build a multiplicative q-white noise out of a given symmetric multiplicative white noise just by multiplying by the M_t^{-1}-process provided the pre-Hilbert space is $\mathbb{M}$-graded in such a way that $L, \tilde{L}$, W and K are 'compatible' with this grading. In this sense, the class of multiplicative q-white noises, $q \neq 1$, is smaller than the class of symmetric multiplicative white noises. For example, let U_t be a unitary q-white noise with $H = \mathbb{C}$. Then L, W and K are all complex $n \times n$-matrices. There is a $\mu_0 \in \mathbb{M}$ such that $H_{(\mu_0)} = \mathbb{C}$ and $H_{(\mu)} = \{0\}$ for $\mu \neq \mu_0$. If q is not a root of unity W and K must be of main diagonal form and L has zero entries except for the diagonal $\{(k,l) : k - l = \mu_0\}$. If q is a νth root of unity W and K have zero entries except for the diagonals $\{(k,l) : k - l \equiv 0 \mod \nu\}$ and L has zero entries except for the diagonals $\{(k,l) : k - l \equiv \mu_0 \mod \nu\}$. A simple non-trivial example is obtained as follows. Let n be equal to 2 and $H_{(-1)} = \mathbb{C}$. Furthermore, put $W = \mathrm{id}$, $K = 0$ and $L = \begin{pmatrix} 0 & 1 \\ 0 & 0 \end{pmatrix}$. Then

$$U_t \begin{pmatrix} \Gamma_t(q) & 0 \\ 0 & \Gamma_t(q)^2 \end{pmatrix}$$

with U_t the solution of

$$U_t = \mathrm{id} + \int_0^t U_r(L\mathrm{d}A_r^* - L^*\mathrm{d}A_r - \frac{1}{2}LL^*\mathrm{d}r)$$

is an example of a unitary q-white noise.

We conclude with the remark that the method applied in this section also can be used to classify white noise on $\mathcal{U}\langle\mathbb{L}\rangle$ with $\mathbb{L}$ a finite group (Section 1.8, Example 6).

4.5. Cocommutative white noise and infinitely divisible representations of groups and Lie algebras. The formulae for white noise become particularly simple for group-like and for Lie elements; see Prop. 4.1.2 and Prop. 4.2.1. In both cases, the comultiplication is cocommutative. We prove explicite expressions for white noise on $\mathbb{Z}_2$-graded cocommutative $*$-bialgebras. These expressions do not involve quantum stochastic integrals or kernels but are convolution products of the convolution exponential of a creation process, a generalized second quantization process, the convolution exponential of an annihilation process, and the 1-parameter convolution semi-group of states associated with the underlying white noise. This generalizes the situation in the group $*$-bialgebra case where we can write

$$j_t(x) = \mathrm{e}^{A_t^*(\eta(x))}\Gamma_t(\rho(x))\mathrm{e}^{A_t(\eta(x^{-1}))}\varphi_t(x),$$

and the convolution product is just the ordinary composition of operators. As a consequence we have the result that the GNS-representation of the infinitely divisible state $\varphi = \varphi_1$ embedded into the 1-parameter convolution semi-group φ_t can be realized on Bose/Fermi Fock space. This generalizes the algebraic part of the so-called Araki-Woods embedding theorem for infinitely divisible positive definite functions on groups [8, 9, 32, 63, 85]; see also [71]. Moreover, it gives a classification of the infinitely divisible representations of $\mathbb{Z}_2$-graded Lie $*$-algebras; cf. [56, 86] where these representations are introduced.

We begin with some general considerations. Let C be a double-module coalgebra such that C is a $\mathbb{C}\mathbb{L}$-comodule and a $\mathbb{C}\mathbb{L}$-module coalgebra and such that the compatibility condition $\mathrm{d}(\lambda.c) = \lambda\mathrm{d}(c)\lambda^{-1}$ is satisfied. The symmetrization C^γ of C is the coalgebra $C \otimes \mathbb{C}\mathbb{L}$ with

$$\Delta_\gamma(c\lambda) = \sum c_{(1)}\mathrm{d}(c_{(2)})\lambda \otimes c_{(2)}\lambda$$

and

$$\varepsilon_\gamma(c\lambda) = \varepsilon(c)$$

where we wrote $c\lambda$ for $c \otimes \lambda$. Suppose that we are given D, η, ρ, ϑ and ψ as in Section 2.3, and in addition a representation $\alpha : \mathbb{L} \to \mathbf{H}(D)$ of $\mathbb{L}$ on D such that

$$\rho(\lambda.c) = \alpha_\lambda \rho(c)\alpha_\lambda^{-1} \quad (4.5.1)$$

$$\eta(\lambda.c) = \alpha_\lambda \eta(c) \quad (4.5.2)$$

$$\vartheta(\lambda.c) = \alpha_{(\lambda^*)^{-1}}\vartheta(c) \quad (4.5.3)$$

$$\psi(\lambda.c) = \psi(c) \quad (4.5.4)$$

for all $c \in \mathcal{C}$ and $\lambda \in \mathbb{L}$. We then can extend η, ρ, ϑ and ψ to $\mathcal{C}^\gamma$ by setting

$$\begin{aligned}\eta_\gamma(c\lambda) &= \eta(c),\\ \rho_\gamma(c\lambda) &= \rho(c)\alpha_\lambda,\\ \vartheta_\gamma(c\lambda) &= \alpha_{\lambda*}\vartheta(c),\\ \psi_\gamma(c\lambda) &= \psi(c).\end{aligned}$$

Next we apply Theorem 2.5.1 to $(\eta_\gamma, \rho_\gamma, \vartheta_\gamma, \psi_\gamma)$ and $\mathcal{C}^\gamma$ and obtain families $(j_{st}^{(\alpha,\gamma)}(c)) \in \underline{\mathcal{A}}_D$ satisfying

$$\begin{aligned} j_{st}^{(\alpha,\gamma)}(c\lambda) = \epsilon(c)\mathrm{id} + \int_s^t j_{sr}^{(\alpha,\gamma)}(c_{(1)}\mathrm{d}(c_{(2)})\lambda) &\\ (\mathrm{d}A_r^*(\eta(c_{(2)})) + \mathrm{d}\Lambda_r(\rho(c_{(2)})\alpha_\lambda - \epsilon(c_{(2)})\mathrm{id}) &\\ + \mathrm{d}A_r(\alpha_{\lambda*}\vartheta(c_{(2)})) + \psi(c_{(2)})\mathrm{d}r). & \qquad (4.5.5)\end{aligned}$$

Using Ito's formula and the conditions on η, ρ, ϑ and ψ, one can show that

$$c\lambda \mapsto j_{st}^{(\alpha,\gamma)}(ce)\Gamma_{st}(\alpha_\lambda)$$

also satisfy (4.5.5) and by the uniqueness of the solution of (4.5.5) in $\underline{\mathcal{A}}_D$ we have

$$j_{st}^{(\alpha,\gamma)}(c\lambda) = j_{st}^{(\alpha,\gamma)}(ce)\Gamma_{st}(\alpha_\lambda).$$

Similarly,

$$j_{st}^{(\alpha,\gamma)}((c.\lambda^{-1})\lambda) = \Gamma_{st}(\alpha_\lambda) j_{st}^{(\alpha,\gamma)}(ce)$$

and, therefore,

$$\Gamma_{st}(\alpha_\lambda) j_{st}^{(\alpha,\gamma)}(ce) = j_{st}^{(\alpha,\gamma)}((\lambda.c)e)\Gamma_{st}(\alpha_\lambda). \qquad (4.5.6)$$

Hence we have for $j_{st}(c) = \Gamma_s(\alpha_{\mathrm{d}(c)}) j_{st}^{(\alpha,\gamma)}(c)$

$$j_{st}(c) = \epsilon(c)\mathrm{id} + \int_s^t (j_{sr} \star \mathrm{d}I_r)(c) \qquad (4.5.7)$$

with

$$\mathrm{d}I_t(c) = \Gamma_t(\alpha_{\mathrm{d}(c)})\mathrm{d}\tilde{I}_t(c)$$

and

$$\tilde{I}_t(c) = A_t^*(\eta(c)) + \Lambda_t(\rho(c) - \epsilon(c)\mathrm{id}) + A_t(\vartheta(c)) + \psi(c)t.$$

In the sequel we assume that $\mathcal{C}$ is a $\mathbb{Z}_2$-graded cocommutative coalgebra. We also assume that D is a $\mathbb{Z}_2$-graded pre-Hilbert space and that η, ρ, ϑ and ψ are even mappings. If we put $\alpha(1)$ equal to the parity operator, i.e. $\alpha(1)\lceil D_{(0)} = \mathrm{id}$ and $\alpha(1)\lceil D_{(1)} = -\mathrm{id}$, the tuple $(\eta, \rho, \vartheta, \psi, \alpha)$ satisfies conditions (4.5.1)-(4.5.4). We put $j_{st}^\eta = j_{st}^{(\eta,0,0,0)}$, $j_{st}^\rho = j_{st}^{(0,\rho,0,0)}$ and $j_{st}^\vartheta = j_{st}^{(0,0,\vartheta,0)}$.

4.5.1. PROPOSITION. *The equation*

$$j_{st}^{(\eta,\rho,\vartheta,\psi)} = j_{st}^{\eta} \star j_{st}^{\rho} \star j_{st}^{\vartheta} \star \varphi_{t-s} \tag{4.5.8}$$

holds for all $(s,t) \in T$.

PROOF: We show that $j_{st}^{\eta} \star j_{st}^{\rho} \star j_{st}^{\vartheta} \star \varphi_{t-s}$ satisfies equation (4.5.7). By quantum Ito's formula

$$\begin{aligned}
&(j_{st}^{\eta} \star j_{st}^{\rho} \star j_{st}^{\vartheta} \star \varphi_{t-s})(c) - \epsilon(c)\mathrm{id} \\
&= \int_s^t \sum (j_{sr}(c_{(1)})\Gamma_r(\alpha_{\mathrm{d}(c_{(2)})}) j_{sr}^{\rho}(c_{(3)}) j_{sr}^{\vartheta}(c_{(4)})\varphi_{r-s}(c_{(5)}) \mathrm{d}A_r^*(\eta(c_{(2)})) \\
&\quad + j_{sr}^{\eta}(c_{(1)}) j_{sr}^{\rho}(c_{(2)})\Gamma_r(\alpha_{\mathrm{d}(c_{(3)})}) j_{sr}^{\vartheta}(c_{(4)})\varphi_{r-s}(c_{(5)}) \mathrm{d}\Lambda_r((\rho - \epsilon\mathrm{id})(c_{(3)})) \\
&\quad + j_{sr}^{\eta}(c_{(1)}) j_{sr}^{\rho}(c_{(2)}) j_{sr}^{\vartheta}(c_{(3)})\Gamma_r(\alpha_{\mathrm{d}(c_{(4)})})\varphi_{r-s}(c_{(5)}) \mathrm{d}A_r^*(\vartheta(c_{(4)})) \\
&\quad + j_{sr}^{\eta}(c_{(1)}) j_{sr}^{\rho}(c_{(2)}) j_{sr}^{\vartheta}(c_{(3)})\varphi_{r-s}(c_{(4)})\Gamma_r(\alpha_{\mathrm{d}(c_{(5)})})\psi(c_{(5)}) \mathrm{d}r)
\end{aligned}$$

which by the commutation rule (4.5.6) and by cocommutativity equals

$$\int_s^t \sum (j_{sr}^{\eta} \star j_{sr}^{\rho} \star j_{sr}^{\vartheta} \star \varphi_{r-s} \star \mathrm{d}I_r)(c).$$

Now equation (4.5.8) follows from the uniqueness of the solution of (4.5.7) in $\underline{A}_D$.◇

For $c \in \mathcal{C}$ and $(s,t) \in T$ we put

$$\mathrm{b}_{st}^{\eta}(c) = \int_s^t \gamma_r(\alpha_{\mathrm{d}(c)}) \mathrm{d}a_r^*(\eta(c))$$

and

$$\mathrm{b}_{st}^{\vartheta}(c) = \int_s^t \gamma_r(\alpha_{\mathrm{d}(c)}) \mathrm{d}a_r(\vartheta(c)).$$

4.5.2. PROPOSITION. *For* $c \in \mathcal{C}$, $\sigma, \tau, \varrho \in \mathcal{S}$ *and* $n \in \mathbb{N}$

$$(\mathrm{b}_{st}^{\eta})^{\star n}(c)(\sigma,\tau,\varrho) = 0 \tag{4.5.9}$$

unless $\#\sigma = n$ *and* $\#\varrho = 0$. *Moreover,*

$$k_{st}^{\eta}(c)(\sigma,\tau,\varrho) = \frac{1}{\sigma!}(\mathrm{b}_{st}^{\eta})^{\star\sigma}(c)(\sigma,\tau,\varrho) = \sum_{l=0}^{\infty} \frac{1}{l!}(\mathrm{b}_{st}^{\eta})^{\star l}(c)(\sigma,\tau,\varrho). \tag{4.5.10}$$

PROOF: Equation(4.5.9) holds for $n = 0$ and for $n = 1$, and if it holds for $n \in \mathbb{N}$ then for $\#\sigma \neq n+1$ or $\#\varrho \neq 0$

$$((\mathrm{b}_{st}^{\eta})^{\star n} \star \mathrm{b}_{st}^{\eta})(c)(\sigma,\tau,\varrho) = 0$$

by the definition of the product of kernels. Taking $C = T(D)$, we have by quantum Ito's formula, the commutation rule (4.5.6) and by cocommutativity

$$\begin{aligned}
&(\mathrm{b}^\eta_{st})^{\star n}(c) \\
&= \int_s^t \sum (\gamma_r(\alpha_{\mathrm{d}(c_{(1)})}) b^\eta_{sr}(c_{(2)}) \ldots \mathrm{b}^\eta_{st}(c_{(n)}) \mathrm{d}a^*_r(\eta(c_{(1)})) + \ldots \\
&\qquad \cdots + \mathrm{b}^\eta_{sr}(c_{(1)}) \ldots \mathrm{b}^\eta_{sr}(c_{(n-1)}) \gamma_r(\alpha_{\mathrm{d}(c_{(n)})}) \mathrm{d}a^*_r(\eta(c_{(n)}))) \\
&= n \int_s^t (\mathrm{b}^\eta_{sr})^{\star(n-1)}(c_{(1)}) \gamma_r(\alpha_{\mathrm{d}(c_{(2)})}) \mathrm{d}a^*_r(\eta(c_{(2)})).
\end{aligned}$$

For $\#\sigma \leq 1$ equation (4.5.10) holds and if it holds for $\#\sigma \leq n$ we have for $\#\sigma = n+1$

$$\begin{aligned}
\sum_{l=0}^{\infty} \frac{1}{l!} (\mathrm{b}^\eta_{st})^{\star l}(c)(\sigma, \tau, \varrho) &= \frac{1}{(n+1)!} (\mathrm{b}^\eta_{st})^{\star(n+1)}(c)(\sigma, \tau, \varrho) \\
&= \frac{n!(n+1)}{(n+1)!} \Big(\int_s^t k^\eta_{sr}(c_{(1)}) \gamma_r(\alpha_{\mathrm{d}(c_{(2)})}) \mathrm{d}a^*_r(\eta(c_{(2)})) \Big)(\sigma, \tau, \varrho) \\
&= k^\eta_{st}(c)(\sigma, \tau, \varrho)
\end{aligned}$$

where we used the previous equality and the equation satisfied by k^η_{st}.◇

We have the analogous result for $\mathrm{b}^\vartheta_{st}$ and k^ϑ_{st}. However, the argument does not work for k^ρ_{st} because $(\mathrm{d}\Lambda)^2$ is not equal to 0. Nevertheless, the k^ρ_{st} can be described very easily. They turn out to be a generalization of the second quantization processes.

For $\mathbb{Z}_2$-graded vector spaces V_1, V_2, W_1 and W_2 and for $R \in \mathbf{L}(V_1, V_2)$ and $S \in \mathbf{L}(W_1, W_2)$ we denote by $(R \otimes S)^\wedge = R \hat{\otimes} S$ the element in $\mathbf{L}(V_1 \otimes W_1, V_2 \otimes W_2)$ given by

$$(R \otimes S)^\wedge (v \otimes w) = (-1)^{\mathrm{d}_S \mathrm{d}_v} (Rv) \otimes (Sw).$$

$\wedge$-Tensor products of length ≥ 3 are defined by

$$(R_1 \otimes \ldots \otimes R_n)^\wedge = (R_1 \otimes \ldots \otimes R_{n-1})^\wedge \hat{\otimes} R_n.$$

4.5.3. Proposition. *For $F \in \mathcal{E}_D$ and $\omega \in \mathcal{S} \cap [s,t)$*

$$(j^\rho_{st}(c)F)(\omega) = (\rho^{\otimes\omega} \circ \Delta_\omega)(c)^\wedge F(\omega).$$

Proof: By Lemma 4.1.1 we have for $\omega = \{t_1 < \cdots < t_n\} \subset [s,t)$ and $F(\omega) = \xi_1 \otimes \ldots \otimes \xi_n$

$$\begin{aligned}
&(j^\rho_{st}(c)F)(\omega) \\
&= \sum_{\tau \subset \omega} ((\rho_\gamma - \epsilon\, \mathrm{id})^{\otimes\tau} \circ (\Delta_\gamma)_\tau)(c)^\omega (\xi_1 \otimes \ldots \otimes \xi_n) \\
&= (\rho_\gamma^{\otimes\omega} \circ (\Delta_\gamma)_\omega)(c)(\xi_1 \otimes \ldots \otimes \xi_n)
\end{aligned}$$

which equals

$$\sum(\rho_\gamma(c_{(1)}d(c_{(2)}\dots c_{(n)}))\xi_1)\otimes(\rho_\gamma(c_{(2)}d(c_{(3)}\dots c_{(n)}))\xi_2)\otimes\dots\otimes(\rho_\gamma(c_{(n)})\xi_n)$$
$$=\sum\prod_{k<l}(-1)^{d(\xi_k)d(c_{(l)})}(\rho(c_{(1)})\xi_1)\otimes\dots\otimes(\rho(c_{(n)})\xi_n)$$
$$=\sum(\rho(c_{(1)})\otimes\dots\otimes\rho(c_{(n)}))^\wedge(\xi_1\otimes\dots\otimes\xi_n).\diamond$$

It follows that for $\omega\in\mathcal{S}$, $\omega=\omega_1\cup\omega_2\cup\omega_3$, $\omega_1\subset[0,s)$, $\omega_2\subset[s,t)$, $\omega_3\subset[t,\infty)$, and $F(\omega)=F_1(\omega_1)\otimes F_2(\omega_2)\otimes F_3(\omega_3)$, $F_i(\omega_i)\in D^{\otimes\omega_i}$, we have

$$(j^\rho_{st}(c)F)(\omega)=((\alpha_{d(c)})^{\otimes\omega_1}F_1(\omega_1))\otimes((\rho^{\otimes\omega}\circ\Delta_\omega)(c)^\wedge F_2(\omega_2))\otimes F_3(\omega_3).$$

For the kernels $k^\eta_{st;m}:=\sum_{l=0}^m\frac{1}{l!}(\mathrm{b}^\eta_{st})^{\star l}$ we have that $k^\eta_{st;m}(\sigma,\tau,\varrho)$ vanishes for $m<\#\sigma$ and equals $k_{st}(\sigma,\tau,\varrho)$ if $m\geq\#\sigma$, so that by Prop. 2.2.2 $\underline{k}^\eta_{st;m}$ converge to j^η_{st} strongly on $\mathcal{E}_D$ for $m\to\infty$. For the same reasons the operators given by the kernels $\sum_{l=0}^m\frac{1}{l!}(\mathrm{b}^\vartheta_{st})^{\star l}$ converge to j^ϑ_{st}. We put $B^\eta_{st}=\underline{\mathrm{b}}^\eta_{st}$ and $B^\vartheta_{st}=\underline{\mathrm{b}}^\vartheta_{st}$ and have

$$j^{(\eta,\rho,\vartheta,\psi)}_{st}=(\exp_\star B^\eta_{st})\star j^\rho_{st}\star(\exp_\star B^\vartheta_{st})\star\varphi_{t-s}$$

where the convolution exponential series are to be understood in the sense of strong convergence on $\mathcal{E}_D$.

We look for linear subspaces of $\mathcal{E}_D$ which are left invariant by the $j_{01}(c)$, $c\in\mathcal{C}$, and which contain the vacuum vector. The reason why we are interested in these subspaces lies in the fact that, in the case when the coalgebra is a $\mathbb{Z}_2$-graded $*$-bialgebra $\mathcal{B}$, ψ is a generator on $\mathcal{B}$ and $\eta=\eta_\psi,\rho=\rho_\psi$, $\vartheta=\tilde\eta$, we can embed the GNS-representation of the state $\varphi=\exp_\star\psi$ into the $*$-representation j_{01} of $\mathcal{B}$ by identifying the cyclic vector of the GNS-representation with the Fock vacuum. Remember that the elements of $\mathcal{F}(L^2(\mathbb{R}_+,D))$ can be regarded as sequences $(F_n)_{n\in\mathbb{N}}$ with $F_n\in L^2(\mathcal{S}_n,D^{\otimes n})$. The figure

$$\begin{array}{ccccccccc}\mathcal{S}&=&\mathcal{S}_0&\dot\cup&\mathcal{S}_1&\dot\cup&\mathcal{S}_2&\dot\cup&\dots\\&&\downarrow&&\downarrow&&\downarrow&&\\\overline{\mathcal{T}}(D)&=&\mathbb{C}&\oplus&D&\oplus&(D\otimes D)&\oplus&\dots\end{array}$$

shows a connection between $\mathcal{F}(L^2(\mathbb{R}_+,D))$ and the $\mathbb{Z}_2$-graded pre-Hilbert space $\overline{\mathcal{T}}(D)$ consisting of all elements $\otimes_{n\in\mathbb{N}}F^{(n)}$ in the tensor Hilbert space $\overline{\bigoplus}_{n\in\mathbb{N}}H^{\overline{\otimes}n}$ with $F^{(n)}\in D^{\otimes n}$. In fact, $\overline{\mathcal{T}}(D)$ can be identified with the linear subspace of $\mathcal{F}(L^2(\mathbb{R}_+,D))$ formed by the functions $\omega\mapsto F(\omega)$ that vanish outside of $\mathcal{S}\cap[0,1]$ and are constant equal to some $F^{(n)}\in D^{\otimes n}$ on $\mathcal{S}_n\cap[0,1]$, $n\in\mathbb{N}$, via the mapping

$$F\mapsto\bigoplus_{n\in\mathbb{N}}\frac{1}{\sqrt{n!}}F^{(n)}$$

which extends to a unitary mapping Ψ from the Hilbert space formed by all functions in $\mathcal{F}(L^2(\mathbb{R}_+,H))$ with support $\mathcal{S}\cap[0,1)$ and which are constant on $\mathcal{S}\cap[0,1]$ to the tensor Hilbert space over H.

Before we proceed with considerations on invariant subspaces we introduce the 'graded' creation, preservation and annihilation processes. We form the graded tensor pre-Hilbert space

$$\mathcal{T}(D) = \bigoplus_{n \in \mathbb{N}} D^{\otimes n}$$

of D which is a linear subspace of $\overline{\mathcal{T}}(D)$. The symmetric group S_n acts on the n-fold tensor product $D^{\otimes n}$ of D with itself through

$$\pi(\xi_1 \otimes \cdots \otimes \xi_n) = \operatorname{sgn}(\pi; \epsilon_1, \ldots, \epsilon_n)\, \xi_{\pi^{-1}(1)} \otimes \cdots \otimes \xi_{\pi^{-1}(m)}$$

where ξ_l are homogeneous elements of D of degree ϵ_l, and for $\epsilon_l \in \mathbb{Z}_2$ the factor $\operatorname{sgn}(\pi; \epsilon_1, \ldots, \epsilon_n)$ denotes the sign of the permutation of $\#\{l : \epsilon_l = 1\}$ elements derived from

$$\pi = \begin{pmatrix} 1 & \ldots & n \\ \pi(1) & \ldots & \pi(n) \end{pmatrix}$$

by eleminating the numbers l in the upper and lower row for which $\epsilon_l = 0$. We then define the linear operator $\mathcal{P}_D$ on $\mathcal{T}(D)$ by

$$\mathcal{P}_D(\xi_1 \otimes \cdots \otimes \xi_m) = \frac{1}{m!} \sum_{\pi \in S_m} \pi(\xi_1 \otimes \cdots \otimes \xi_m).$$

We define the linear operator

$$\tilde{\Theta} : \mathcal{T}(D) \to \mathcal{T}(D^{(0)}) \otimes \mathcal{T}(D^{(1)})$$

by

$$\tilde{\Theta}(\xi_1 \otimes \cdots \otimes \xi_m) = \binom{m}{k}^{-\frac{1}{2}} \zeta_1 \otimes \cdots \otimes \zeta_k \otimes \zeta_{k+1} \otimes \cdots \otimes \zeta_m$$

where $\zeta_1, \ldots, \zeta_k$ are the even and $\zeta_{k+1}, \ldots, \zeta_m$ are the odd factors of the tensor $\xi_1 \otimes \cdots \otimes \xi_m$. Moreover, define

$$\tilde{\Theta}^\dagger : \mathcal{T}(D^{(0)}) \otimes \mathcal{T}(D^{(1)}) \to \mathcal{T}(D)$$

by

$$\tilde{\Theta}^\dagger(\zeta_1 \otimes \ldots \zeta_k \otimes \zeta_{k+1} \otimes \cdots \otimes \zeta_m) = \sum_{\pi \in S_{m,k}} \binom{m}{k}^{-\frac{1}{2}} \pi(\zeta_1 \otimes \ldots \zeta_k \otimes \zeta_{k+1} \otimes \cdots \otimes \zeta_m)$$

where $S_{m,k}$ denotes the subgroup of S_m formed by the permutations which respect the order of $\{1,\ldots,k\}$ and $\{k+1,\ldots,m\}$. Then

$$\begin{aligned}
(\tilde{\Theta})^* &= \tilde{\Theta}^\dagger, \\
\tilde{\Theta}\circ(\tilde{\Theta})^* &= \mathrm{id}, \\
\mathcal{P}_D &= (\tilde{\Theta})^*\circ(\mathcal{P}_{D^{(0)}}\otimes\mathcal{P}_{D^{(1)}})\circ\tilde{\Theta}, \\
(\mathcal{P}_D)^2 &= \mathcal{P}_D, \\
(\mathcal{P}_D)^* &= \mathcal{P}_D, \\
\tilde{\Theta}\circ\mathcal{P}_D &= (\mathcal{P}_{D^{(0)}}\otimes\mathcal{P}_{D^{(1)}})\circ\tilde{\Theta}.
\end{aligned}$$

We set $\mathcal{S}(D)=\mathcal{P}_D(\mathcal{T}(D))$. The restriction Θ of $\tilde{\Theta}$ to $\mathcal{S}(D)$ is an isomorphism from the pre-Hilbert space $\mathcal{S}(D)$ to the pre-Hilbert space $\mathcal{S}(D^{(0)})\otimes\mathcal{S}(D^{(1)})$. If D is the orthogonal sum $D_1\oplus D_2$ of two graded linear subspaces D_1 and D_2 we have

$$\mathcal{S}(D)\cong\mathcal{S}(D_1)\otimes\mathcal{S}(D_2)$$

where we use the canonical isomorphisms

$$\begin{aligned}
\mathcal{S}(D^{(0)}) &\cong \mathcal{S}(D_1^{(0)})\otimes\mathcal{S}(D_2^{(0)}), \\
\mathcal{S}(D^{(1)}) &\cong \mathcal{S}(D_1^{(1)})\otimes\mathcal{S}(D_2^{(1)});
\end{aligned}$$

see [4, 14].

For $\xi\in D$ define the operators $\tilde{A}^*(\xi)$ and $\tilde{A}(\xi)$ on $\mathcal{T}(D)$ by

$$\begin{aligned}
\tilde{A}^*(\xi)\xi_1\otimes\cdots\otimes\xi_n &= (n+1)^{\frac{1}{2}}\xi\otimes\xi_1\otimes\cdots\otimes\xi_n, \\
\tilde{A}^*(\xi)\Omega &= \xi
\end{aligned}$$

and

$$\begin{aligned}
\tilde{A}(\xi)\xi_1\otimes\cdots\otimes\xi_{n+1} &= (n+1)^{\frac{1}{2}}\langle\xi,\xi_1\rangle\xi_2\otimes\cdots\otimes\xi_{n+1}, \\
\tilde{A}(\xi)\zeta &= \langle\xi,\zeta\rangle\Omega, \\
\tilde{A}(\xi)\Omega &= 0.
\end{aligned}$$

Then $\tilde{A}^*(\xi)=\tilde{A}(\xi)^*$, and we define the *creation operators* $\hat{A}^*(\xi)$ and the *annihilation operators* $\hat{A}(\xi)$ on $\mathcal{S}(D)$ by

$$\hat{A}^*(\xi)=\mathcal{P}_D\circ\tilde{A}^*(\xi)\lceil\mathcal{S}(D)$$

and

$$\hat{A}(\xi)=\tilde{A}(\xi)\lceil\mathcal{S}(D).$$

For $R\in\mathbf{H}(D)$ we define the operator $\tilde{\Lambda}(R)$ on $\mathcal{T}(D)$ by

$$\begin{aligned}
\tilde{\Lambda}(R)\lceil D^{\otimes n} &= R\hat{\otimes}\mathrm{id}\hat{\otimes}\ldots\hat{\otimes}\mathrm{id}+\mathrm{id}\hat{\otimes}R\hat{\otimes}\mathrm{id}\hat{\otimes}\ldots\hat{\otimes}\mathrm{id}+\ldots \\
&\quad\cdots+\mathrm{id}\hat{\otimes}\ldots\hat{\otimes}\mathrm{id}\hat{\otimes}R, \\
\tilde{\Lambda}(R)\Omega &= 0.
\end{aligned}$$

The operators $\hat{\Lambda}(R) = \tilde{\Lambda}(R)\lceil\mathcal{S}(D)$ are called *differential second quantization* or *preservation operators*. The mappings $\xi \mapsto \hat{A}^*(\xi)$ and $R \mapsto \hat{\Lambda}(R)$ are even, linear mappings from D to $\mathbf{H}(\mathcal{S}(D))$ and from $\mathbf{H}(D)$ to $\mathbf{H}(\mathcal{S}(D))$ respectively. The latter also is hermitian. The mapping $\xi \mapsto \hat{A}(\xi)$ is an even, antilinear mapping from D to $\mathbf{H}(\mathcal{S}(D))$. For a decomposition $D = D_1 \oplus D_2$ the canonical isomorphism between $\mathcal{S}(D)$ and $\mathcal{S}(D_1) \otimes \mathcal{S}(D_2)$ gives

$$\begin{aligned}
\hat{A}^*(\xi_1 \oplus \xi_2) &\cong \hat{A}^*(\xi_1)\hat{\otimes}\mathrm{id} + \mathrm{id}\hat{\otimes}\hat{A}^*(\xi_2),\\
\hat{A}(\xi_1 \oplus \xi_2) &\cong \hat{A}(\xi_1)\hat{\otimes}\mathrm{id} + \mathrm{id}\hat{\otimes}\hat{A}(\xi_2),\\
\hat{\Lambda}(R_1 \oplus R_2) &\cong \hat{\Lambda}(R_1)\hat{\otimes}\mathrm{id} + \mathrm{id}\hat{\otimes}\lambda(R_2)
\end{aligned}$$

for $\xi_1 \in D_1$, $\xi_2 \in D_2$, $R_1 \in \mathbf{H}(D_1)$ and $R_2 \in \mathbf{H}(D_2)$. In particular, for $D_1 = D^{(0)}$ and $D_2 = D^{(1)}$ we have

$$\begin{aligned}
\hat{A}^*(\xi) &\cong \hat{A}^*(\xi^{(0)}) \otimes \mathrm{id} + \mathrm{id} \otimes \hat{A}^*(\xi^{(1)}),\\
\hat{A}(\xi) &\cong \hat{A}(\xi^{(0)}) \otimes \mathrm{id} + \mathrm{id} \otimes \hat{A}(\xi^{(1)})
\end{aligned}$$

where $\xi \in D$ and $\xi^{(0)}$ and $\xi^{(1)}$ denote the even and the odd part of ξ respectively. Notice that for an *odd* operator R in $\mathbf{H}(D)$ the operator $\hat{\Lambda}(R)$ does not leave invariant $\mathcal{S}(D^{(0)})$ or $\mathcal{S}(D^{(1)})$. We have the following *canonical commutation relations*

$$\begin{aligned}
[\hat{A}(\xi), \hat{A}(\zeta)] &= [\hat{A}^*(\xi), \hat{A}^*(\zeta)] = 0,\\
[\hat{A}(\xi), \hat{A}^*(\zeta)] &= \langle \xi, \zeta \rangle \mathrm{id},\\
[\hat{\Lambda}(R), \hat{\Lambda}(S)] &= \hat{\Lambda}([R, S]),\\
[\hat{\Lambda}(R), \hat{A}^*(\xi)] &= \hat{A}^*(R\xi),\\
[\hat{\Lambda}(R), \hat{A}(\xi)] &= -\hat{A}(R^*\xi)
\end{aligned}$$

for $\xi, \zeta \in D$ and $R, S \in \mathbf{H}(D)$ where the brackets are to be understood as graded commutators, i.e.

$$[A, B] = AB - (-1)^{\mathrm{d}_A \mathrm{d}_B} BA$$

for $A, B \in \mathbf{L}(\mathcal{S}(D))$. We equip the Hilbert space $L^2(\mathbb{R}_+)$ with the trivial graduation. For $\xi \in D$ and $R \in \mathbf{H}(D)$ we define operators on $\mathcal{S}(L^2(\mathbb{R}_+) \otimes D)$ by

$$\begin{aligned}
\hat{A}^*_{st}(\xi) &= \hat{A}^*(\chi_{[s,t)} \otimes \xi),\\
\hat{A}_{st}(\xi) &= \hat{A}(\chi_{[s,t)} \otimes \xi),\\
\hat{\Lambda}_{st}(R) &= \hat{\Lambda}(\chi_{[s,t)} \otimes R)
\end{aligned}$$

where, in the third definition, we identified $\chi_{[s,t)} \in L^2(\mathbb{R}_+)$ with the corresponding multiplication operator on $L^2(\mathbb{R}_+)$. The operators $\hat{A}^*(\xi)$, $\hat{\Lambda}(R)$ and $\hat{A}(\xi)$ on the completion $\hat{\mathcal{F}}(D)$ of $\mathcal{S}(D)$ are closable and the domains of their closures contain the linear subspace

$$\mathcal{E}_{(D)} = \Big(\bigcap_{\alpha \in \mathbb{R}_+} \mathrm{Dom}\, \alpha^{\hat{\Lambda}(\mathrm{id})} \Big) \cap \bigcup_{E \in \mathcal{P}_0(D)} \overline{\mathcal{T}}(E)$$

of $\hat{\mathcal{F}}(D)$. The closures leave $\mathcal{E}_{(D)}$ invariant and we extend $\hat{A}^*(\xi), \hat{\Lambda}(R)$ and $\hat{A}(\xi)$ to operators on $\mathcal{E}_{(D)}$. We extend $\hat{A}^*_{st}(\xi)$, $\hat{\Lambda}_{st}(R)$ and $\hat{A}_{st}(\xi)$ to operators on $\mathcal{E}_{(L^2(\mathbb{R}_+)\otimes D)}$.

4.5.4. PROPOSITION. *The operators $B^\eta_{01}(c)$ and $B^\vartheta_{01}(c)$ leave $\Psi^{-1}(\mathcal{E}_{(D)})$ invariant and*

$$B^\eta_{01}(c)\lceil\Psi^{-1}(\mathcal{E}_{(D)}) = \Psi^{-1}\circ\hat{A}^*(\eta(c))\circ\Psi,$$
$$B^\vartheta_{01}(c)\lceil\Psi^{-1}(\mathcal{E}_{(D)}) = \Psi^{-1}\circ\hat{A}(\vartheta(c))\circ\Psi.$$

PROOF: Using Lemma 4.1.1 we have for

$$F^{(n)} = \sum_i \xi_{i1}\otimes\ldots\otimes\xi_{in}\in\mathcal{E}_D\cap\Psi^{-1}(\hat{\mathcal{F}}(H)) = \Psi^{-1}(\mathcal{E}_{(D)})$$

that $(B^\eta_{01}(c)F^{(n)})(\omega) = 0$ if $\#\omega\neq n+1$ and for $\#\omega = n+1$

$$\begin{aligned}
&(B^\eta_{01}(c)F^{(n)})(\omega)\\
&=\sum_i\sum_{l=1}^n(-1)^{\mathrm{d}(c)(\mathrm{d}(\xi_1)+\cdots+\mathrm{d}(\xi_{l-1}))}\xi_{i1}\otimes\ldots\otimes\xi_{i,l-1}\otimes\eta(c)\otimes\xi_{il}\otimes\ldots\otimes\xi_{in}\\
&=\sqrt{n+1}\hat{A}^*(\eta(c))(\sum_i\xi_{i1}\otimes\ldots\otimes\xi_{in})\\
&=(\Psi^{-1}\circ\hat{A}^*(\eta(c))\circ\Psi)(F^{(n)}).\diamondsuit
\end{aligned}$$

4.5.5. THEOREM. *The operators $j^{(\eta,\rho,\vartheta,\psi)}_{01}(c)$ leave $\Psi^{-1}(\mathcal{E}_{(D)})$ invariant.*

PROOF: From Prop. 4.5.4 and the fact that $\sum_{l=0}^m\frac{1}{l!}(B^\eta_{01})^{\star l}(c)$ strongly converges to $j^\eta_{01}(c)$ on $\mathcal{E}_D$ we conclude that $j^\eta_{01}(c)$ leaves $\Psi^{-1}(\hat{\mathcal{F}}(H))$ invariant. Since $j^\eta_{01}(c)$ leaves $\mathcal{E}_D$ invariant, we have that $j^\eta_{01}(c)$ leaves $\Psi^{-1}(\mathcal{E}_{(D)})$ invariant. The same argument yields that $j^\vartheta_{01}(c)$ leaves $\Psi^{-1}(\mathcal{E}_{(D)})$ invariant. It remains to show that $(\rho^{\otimes n}\circ\Delta_n)(c)^\wedge F^{(n)}$ is in $\Psi^{-1}(\mathcal{E}_{(D)})$ for $F^{(n)}\in\Psi^{-1}(\mathcal{E}_{(D)})$. This follows from cocommutativity.$\diamondsuit$

We can realize $j^{(\eta,\rho,\vartheta,\psi)}_{01}(c)$ on the Bose/Fermi Fock space $\hat{\mathcal{F}}(H)$. The $j_{01}(c)$ are operators on $\mathcal{E}_{(D)}$ and (with the obvious notation)

$$j_{01} = \exp_\star(\hat{A}^*\circ\eta)\star(\otimes_{n\in\mathbb{N}}\rho^{\otimes n}\circ\Delta_n)^\wedge\star\exp_\star(\hat{A}\circ\vartheta)\star\varphi. \tag{4.5.11}$$

We mention that, if $\mathcal{B}$ is a $\mathbb{Z}_2$-graded $*$-bialgebra and ψ is a generator on $\mathcal{B}$, a white noise with generator ψ can be realized on the Bose/Fermi Fock space $\hat{\mathcal{F}}(L^2(\mathbb{R}_+,H))$ and

$$j_{st} = \exp_\star(\hat{A}^*_{st}\circ\eta)\star(\otimes_{n\in\mathbb{N}}\rho^{\otimes n}_{st}\circ\Delta_n)^\wedge\star(\exp_\star\hat{A}_{st}\circ\tilde{\eta})\star\varphi_{t-s}$$

with $\rho_{st} = \lambda_{[s,t)}\otimes\rho\in\mathbf{H}(L^2(\mathbb{R}_+)\otimes D)$. The domain of $j_{st}(b)$ is $\mathcal{E}_{(L^2(\mathbb{R}_+)\otimes D)}$. This can be proved by checking the axioms (WN1)-(WN4) where cocommutativity must be applied and by using (4.5.11) to see that ψ is the generator.

A state φ on an involutive (γ,α)-bialgebra $\mathcal{B}$ is called *infinitely divisible* if there are states φ_n, $n\in\mathbb{N}$, such that $(\varphi_n)^{\star n} = \varphi$. A state of the form $\exp_\star\psi$ with ψ a generator on $\mathcal{B}$ is infinitely divisible.

We state the following *embedding theorem for infinitely divisible states on cocommutative $*$-bialgebras*

4.5.6. THEOREM. *Let $\mathcal{B}$ be a $\mathbb{Z}_2$-graded cocommutative $*$-bialgebra and let ψ be a generator on $\mathcal{B}$. Then the GNS-representation of $\varphi = \exp_\star \psi$ can be embedded into the $*$-representation j of $\mathcal{B}$ on $\mathcal{E}_{(D)}$ given by*

$$j = \exp_\star(\hat{A}^* \circ \eta_\psi) \star (\otimes_{n\in\mathbb{N}} \rho_\psi^{\otimes n} \circ \Delta_n)^\wedge \star \exp_\star(\hat{A} \circ \tilde{\eta}_\psi) \star \varphi$$

via the intertwining mapping

$$\pi_\varphi(b)\Omega_\varphi \mapsto j_{01}(b)\Omega$$

which extends to a unitary mapping from H_φ to the closure of $\{j_{01}(b)\Omega : b \in \mathcal{B}\}$ in $\hat{\mathcal{F}}(H)$. Here $(\pi_\varphi, H_\varphi, \Omega_\varphi)$ denotes the GNS-triplet associated with φ.$\diamondsuit$

For a $\mathbb{Z}_2$-graded algebra $\mathcal{A}$ denote by $\mathcal{A}_\mathbf{g}$ the $\mathbb{Z}_2$-graded Lie algebra derived from $\mathcal{A}$ by defining the Lie bracket by

$$[a, b] = ab - (-1)^{\mathrm{d}_a \mathrm{d}_b} ba,$$

$a, b \in \mathcal{A}$. For a $\mathbb{Z}_2$-graded vector space V we denote by $\mathbf{g}(V)$ the smallest $\mathbb{Z}_2$-graded Lie subalgebra of $\mathcal{T}(V)_\mathbf{g}$ containing V. We are now ready to state our result on the GNS-representation given by an infinitely divisible state on a tensor $*$-algebra.

4.5.7. THEOREM. *Let V be a $\mathbb{Z}_2$-graded involutive vector space and let φ be an even, infinitely divisible state on $\mathcal{T}(V)$. Then there exist a $\mathbb{Z}_2$-graded pre-Hilbert space D, an even linear mapping $\eta_0 : \mathcal{B} \to D$, an even, hermitian linear mapping $\rho_0 : \mathcal{B} \to \mathbf{H}(D)$ with*

$$D = \{\rho_0(v_1)\dots\rho_0(v_n)\eta_0(v_{n+1}) : n \in \mathbb{N}, v_1, \dots, v_{n+1} \in V\}$$

and an even, hermitian linear functional ψ_0 on V such that the GNS-representation of φ can be embedded into the even $*$-representation j of $\mathcal{T}(V)$ on $\mathcal{E}_{(D)}$ given by

$$j(b) = \hat{A}^*(\eta(b)) + \hat{\Lambda}(\rho(b)) + \hat{A}(\eta(b^*)) + \psi(b)\mathrm{id} \tag{4.5.12}$$

for $b \in V$, i.e.

$$\varphi(b) = \langle \Omega, j(b)\Omega \rangle.$$

Here η, ρ and ψ are given by (4.2.12)-(4.2.14) and $(D, \eta_0, \rho_0, \psi_0)$ is unique up to equivalence. Moreover, $\varphi = \exp_\star \psi$ and (4.5.12) holds for all $b \in \mathbf{g}(V)$.

PROOF: If φ is an even, infinitely divisible state then there exists a unique generator ψ such that $\varphi = \exp_\star \psi$; see [33, 56, 86] and Section 5.2. Now apply Theorem 4.5.6. By the Poincaré-Birkhoff-Witt theorem for $\mathbb{Z}_2$-graded Lie algebras (see [46]) b lies in $\mathbf{g}(V)$ if and only if b is a Lie element, so that (4.5.12) holds for all $b \in \mathbf{g}(V)$.$\diamondsuit$

Finally, we state the extra conditions on η_0, ρ_0 and ψ_0 for ψ to vanish on the ideal $\mathcal{I}_\mathbf{g}$ in $\mathcal{T}(\mathbf{g})$ generated by

$$\{u \otimes v - (-1)^{\mathrm{d}_u \mathrm{d}_v} v \otimes u - [u, v] : u, v \in \mathbf{g}\}$$

in the case when g is a $\mathbb{Z}_2$-graded Lie $*$-algebra. One easily checks that $\psi\lceil\mathcal{I}_{\mathrm{g}} = 0$ if and only if

$$\rho_0([u,v]) = [\rho_0(u), \rho_0(v)] \tag{4.5.13}$$

$$\eta_0([u,v]) = \rho_0(u)\eta_0(v) - (-1)^{\mathrm{d}_u\mathrm{d}_v}\rho_0(v)\eta_0(u) \tag{4.5.14}$$

$$\psi_0([u,v]) = \langle\eta_0(u^*), \eta_0(v)\rangle - (-1)^{\mathrm{d}_u\mathrm{d}_v}\langle\eta_0(v^*), \eta_0(u)\rangle \tag{4.5.15}$$

are satisfied for all $u, v \in$ g. Thus ρ_0 has to be a $*$-representation of g. Condition (4.5.14) means that η_0 is a 1-cocycle with respect to ρ_0 and condition (4.5.15) says that the left hand side is the coboundary of ψ_0 in the terminology of cohomology theory for Lie algebras; see e.g. [45]. We conclude that the GNS-representation of even, infinitely divisible states on $\mathcal{T}(\mathrm{g})$ (these representation were called 'infinitely divisible representations of Lie algebras' in [56, 86]) can be embedded into representations of the form (4.5.12) with η_0, ρ_0 and ψ_0 satisfying the additional conditions (4.5.13)-(4.5.15).

5. Quadratic components of conditionally positive linear functionals.

Let G be a compact Lie group. As a consequence of Hunt's formula [44] a generator ψ on $\mathcal{R}_0(G)$ always can be decomposed into a 'maximal quadratic' part and a part with no 'quadratic component' in the following sense. One has

$$\psi(f) = \sum_{k=1}^{n} \alpha_k \mathrm{X}_k(f)(e) + \sum_{k,l=1}^{n} \alpha_{k,l} \mathrm{X}_k \mathrm{X}_l(f)(e) + r(f)$$

where $\mathrm{X}_1, \ldots, \mathrm{X}_n$ is a basis of the Lie algebra g of G, $\alpha_1, \ldots, \alpha_n$ are real numbers, $(\alpha_{kl})_{k,l=1,\ldots,n}$ is a positive semi-definite real $n \times n$-matrix, and r is another generator which cannot be decomposed into

$$f \mapsto \mathrm{Y}^2(f)(e) + r'(f)$$

with $\mathrm{Y} \in \mathrm{g}$, $\mathrm{Y} \neq 0$, and r' a generator. In the sense of this chapter, $\psi - r$ is a maximal quadratic component of ψ. It is uniquely determined by ψ up to adding a term of the form $f \mapsto \mathrm{Y}(f)(e)$ with $\mathrm{Y} \in \mathrm{g}$.

We generalize this result. First we say what we mean by a quadratic linear functional on a $*$-bialgebra. Then we find a cohomological condition on the $*$-bialgebra which guarantees the existence of maximal quadratic components. We show that commutative $*$-bialgebras satisfy this condition which from the point of view of *-bialgebras explains why Hunt's theorem holds. Next we prove that $\mathcal{U}\langle n\rangle$ has this property, too. This is remarkable, because this $*$-bialgebra is neither commutative nor cocommutative. We apply the fact that free algebras always satisfy our condition to prove that an infinitely divisible state on the formal algebra of the canonical commutation relations can be written in a unique way as the convolution of a gaussian (i.e. quasi free) state and a 'classical' state with no gaussian component. Moreover, our result contains the result of [56] which says that an infinitely divisible state on the algebra of the canonical anti-commutation relations must be gaussian.

5.1. Maximal quadratic components.

As we do not need the comultiplication for most of this section, we start from a more consisting of a $*$-algebra $\mathcal{B}$ and a $*$-algebra homomorphism ϵ from $\mathcal{B}$ to the field of complex numbers. The notion of conditional positivity and the construction of D_ψ, η_ψ and ρ_ψ out of a conditionally positive linear functional ψ on $\mathcal{B}$ are the same as in Section 2.3. For $n \in \mathbb{N}$ denote by K^n the linear span of the set $\{a_1 \ldots a_n : a_j \in \text{Kern } \epsilon\}$. We call a conditionally positive hermitian linear functional on $\mathcal{B}$ *quadratic* if it satisfies one of the conditions of the following proposition which is a straightforward verification.

5.1.1. Proposition. *For a conditionally positive, hermitian linear functional ψ on $\mathcal{B}$ the following are equivalent*

(i) $\psi \lceil K^3 = 0$
(ii) $\psi(a^* a) = 0$ *for all* $a \in K^2$
(iii)

$$\begin{aligned}\psi(abc) &= \psi(ab)\epsilon(c) + \psi(ac)\epsilon(b) + \epsilon(a)\psi(bc) \\ &\quad - \epsilon(a)\epsilon(b)\psi(c) - \epsilon(a)\psi(b)\epsilon(c) - \psi(a)\epsilon(b)\epsilon(c)\end{aligned}$$

for all $a, b, c \in \mathcal{B}$.

(iv) $\rho_\psi\lceil K^1 = 0$

(v) $\rho_\psi(a) = \epsilon(a)\mathrm{id}$ *for all* $a \in \mathcal{B}$

(vi) $\eta_\psi\lceil K^2 = 0$

(vii) $\eta_\psi(ab) = \epsilon(a)\eta_\psi(b) + \eta_\psi(a)\epsilon(b)$ *for all* $a, b \in \mathcal{B}$.◇

We say that ψ is *degenerate* quadratic if ψ is hermitian and vanishes on K^2. For convenience, we use the following cohomological notation; cf. Section 2.3. Given a $*$-representation ρ of $\mathcal{B}$ on a pre-Hilbert space D we turn D into a $\mathcal{B}$-bimodule by setting

$$aub = \rho(a)u\epsilon(b)$$

for $a, b \in \mathcal{B}$ and $u \in D$. We denote by $Z^n(D, \rho)$ the n-cocycles and by $B^n(D, \rho)$ the n-coboundaries of the corresponding Hochschildt cohomology theory. Condition (vii) of Proposition 5.1.1 says that $\eta_\psi \in Z^1(D_\psi, \epsilon)$. It is clear that the bilinear form $\mathcal{L}(\psi)$ on $\mathcal{B}$ with

$$\mathcal{L}(\psi)(a, b) = -\langle \eta_\psi(a^*), \eta_\psi(b)\rangle = -(a^*, b)_\psi$$

is the coboundary of ψ (we write $\partial\psi = -\mathcal{L}(\psi)$) which yields $\mathcal{L}(\psi) \in B^2(\mathbb{C}, \epsilon)$. If $\mathcal{L}$ is a bilinear form on $\mathcal{B}$ satisfying $\mathcal{L}(a, b) = \overline{\mathcal{L}(b^*, a^*)}$ then we have that $\partial\psi = \mathcal{L}$ implies $\partial\psi^* = \mathcal{L}$ where $\psi^* \in \mathcal{B}^*$ with $\psi^*(a) = \overline{\psi(a^*)}$. It follows that $\partial(\frac{1}{2}(\psi + \psi^*)) = \mathcal{L}$ which shows that a solution ψ of $\partial\psi = \mathcal{L}$ can assumed to be hermitian without loss of generality. By the following simple proposition the study of quadratic linear functionals can be reduced to the study of elements of $Z^1(\mathbb{C}, \epsilon)$ in many cases.

5.1.2.PROPOSITION. *Let* $\mathcal{B}$ *be finitely generated and let* η *be in* $Z^1(D, \epsilon)$. *Then there exist* $m \in \mathbb{N}$ *and* $\mathrm{d}_1, \ldots, \mathrm{d}_m$ *in* $Z^1(\mathbb{C}, \epsilon)$ *such that*

$$\langle \eta(a), \eta(b)\rangle = \sum_{i=1}^m \overline{\mathrm{d}_i(a)}\mathrm{d}_i(b) \tag{5.1.1}$$

for all $a, b \in \mathcal{B}$.

PROOF: We claim that $\eta(\mathcal{B})$ is finite-dimensional. For let $\{a_0, a_1, \ldots, a_k\}$ be a set of generators of $\mathcal{B}$. We can assume that $a_0 = 1$ and $a_j \in K^1$ for $j \geq 1$, otherwise we pass to $a_j - \epsilon(a_j)1$. Since $\eta \in Z^1(D, \epsilon)$ means $\eta(a) = 0$ for $a \in K^2$ and since $\eta(1) = 0$, we have $\eta(\mathcal{B}) = \mathrm{lin}\{\eta(a_1), \ldots, \eta(a_k)\}$. Now choose an orthonormal basis $\{e_1, \ldots, e_m\}$ of $\eta(\mathcal{B})$. Then $\mathrm{d}_i \in Z^1(\mathbb{C}, \epsilon)$ where $\mathrm{d}_i(a) = \langle e_i, \eta(a)\rangle$, and (5.1.1) holds.◇

5.1.3.PROPOSITION. *Let* $\mathcal{B}$ *be finitely generated and let* ψ *be a quadratic linear functional on* $\mathcal{B}$. *Then there exist* $m \in \mathbb{N}$ *and* $\mathrm{d}_1, \ldots, \mathrm{d}_m$ *in* $Z^1(\mathbb{C}, \epsilon)$ *such that*

$$\psi(ab) = \psi(a)\epsilon(b) + \epsilon(a)\psi(b) + \sum_{i=1}^m \mathrm{d}_i^*(a)\mathrm{d}_i(b)$$

for all $a, b \in \mathcal{B}$. If $\mathcal{B}$ is commutative, the d_i can be chosen to be hermitian.

PROOF: The first part is only an application of Prop. 5.1.2 to η_ψ in the set $Z^1(D_\psi, \epsilon)$.- For the second part one only has to remark that, if $\mathcal{B}$ is commutative,

$$\psi(ab) = \psi(a)\epsilon(b) + \epsilon(a)\psi(b) + \sum_{i=1}^{m}(\mathrm{d}_i^{(1)}(a)\mathrm{d}_i^{(1)}(b) + \mathrm{d}_i^{(2)}(a)\mathrm{d}_i^{(2)}(b))$$

with

$$\mathrm{d}_i^{(1)} = \frac{1}{2}(\mathrm{d}_i + (\mathrm{d}_i)^*),$$
$$\mathrm{d}_i^{(2)} = \frac{i}{2}(\mathrm{d}_i - (\mathrm{d}_i)^*).\diamond$$

Assume for a moment that $\mathcal{B}$ has a comultiplication Δ and a counit ϵ turning $\mathcal{B}$ into a symmetric $*$-bialgebra. A *left invariant derivation* D on $\mathcal{B}$ is a derivation on the algebra $\mathcal{B}$ such that

$$\mathrm{D} = (\mathrm{id} \otimes (\epsilon \circ \mathrm{D})) \circ \Delta.$$

If D is a derivation then $\epsilon \circ \mathrm{D} \in Z^1(\mathbb{C}, \epsilon)$. If, on the other hand, $\mathrm{d} \in Z^1(\mathbb{C}, \epsilon)$ then $\mathrm{D} = (\mathrm{id} \otimes \mathrm{d}) \circ \Delta$ is a left invariant derivation. This establishes a 1-1-correspondence between left invariant derivations on $\mathcal{B}$ and elements in $Z^1(\mathbb{C}, \epsilon)$. The hermitian elements in $Z^1(\mathbb{C}, \epsilon)$ are precisely the degenerate quadratic linear functionals.

If $\mathcal{B}$ is commutative and finitely generated as an algebra then, by Prop. 5.1.3, for a quadratic ψ there exist $m \in \mathbb{N}$ and hermitian left invariant derivations $\mathrm{D}_1, \ldots, \mathrm{D}_m$ on $\mathcal{B}$ such that

$$\psi - \frac{1}{2}\sum_{i=1}^{m}(\epsilon \circ (\mathrm{D}_i)^2)$$

is a hermitian left invariant derivation. The proof of the following is again straightforward.

5.1.4.PROPOSITION. *For a linear functional ψ on $\mathcal{B}$ the following are equivalent*

(i) $\psi \in Z^1(\mathbb{C}, \epsilon)$
(ii) $\psi\lceil K^2 = 0$
(iii) $\psi(a^*a) = 0$ *for all* $a \in K^1$
(iv) ψ *is conditionally positive and* $D_\psi = \{0\}.\diamond$

Next we investigate under which circumstances we can extract a maximal quadratic component from a given hermitian conditionally positive linear functional. All is based on the following simple considerations. Let ρ be a $*$-representation of $\mathcal{B}$ on a pre-Hilbert space D and denote by H the completion of D. We form the closed subspace

$$H_1 = \{u \in H : \langle u, \rho(a)v\rangle = \langle u, v\rangle\epsilon(a) \text{ for all } a \in \mathcal{B} \text{ and for all } v \in D\}$$

of H. We denote by P_1 the orthogonal projection onto H_1 and we set $H_2 = (H_1)^\perp$, $P_2 = \mathrm{id} - P_1$, $D_1 = P_1 D$ and $D_2 = P_2 D$.

5.1.5. LEMMA. *For a $*$-representation ρ of $\mathcal{B}$ the equation*

$$\rho_2(a)P_2v = \rho(a)v - \epsilon(a)P_1v,$$

$v \in D$, $a \in \mathcal{B}$, defines a $*$-representation of $\mathcal{B}$ on D_2, and

$$\rho_2(a)P_2v = P_2\rho(a)v.$$

Moreover, $\epsilon \mathrm{id} \oplus \rho_2$ is a $*$-representation of $\mathcal{B}$ which is an extension of ρ, i.e.

$$D \subset D_1 \oplus D_2$$

and

$$(\epsilon\,\mathrm{id} \oplus \rho_2)\lceil D = \rho.$$

Given $\eta \in Z^1(D,\rho)$ we have that

$$P_1 \circ \eta \in Z^1(D_1,\epsilon)$$

and

$$P_2 \circ \eta \in Z^1(D_2,\rho_2).$$

PROOF: First we must prove that $\rho_2(a) : D_2 \to H$ is a well-defined linear mapping. So assume $P_2v = 0$. Then

$$\begin{aligned}\rho_2(a)P_2v &= \rho(a)v - \epsilon(a)P_1v \\ &= \rho(a)P_1v - \epsilon(a)P_1v \\ &= 0,\end{aligned}$$

because for $u \in D$

$$\langle \rho(a)P_1v - \epsilon(a)P_1v, u\rangle = \langle P_1v, \rho(a^*)u - \epsilon(a^*)u\rangle = 0.$$

Next we have

$$\langle u, \rho_2(a)P_2v - P_2\rho(a)v\rangle = \langle u, P_1\rho(a)v - \epsilon(a)P_1v\rangle = 0$$

which proves $\rho_2(a)P_2v = P_2\rho(a)v$. Thus $\rho_2(a)$ maps D_2 to D_2 and the remaining part is straightforward.◇

In the sequel, for a given $\eta \in Z^1(D,\rho)$ we put $\eta_1 = \eta \circ P_1$ and $\eta_2 = \eta \circ P_2$. Notice that H_2 is the closure of $\eta(K^2)$.

Let η be in $Z^1(D,\rho)$. A pair (E,θ) with E a pre-Hilbert space and $\theta \in Z^1(E,\epsilon)$ such that $\theta(\mathcal{B}) = E$ is called a *derivation component* of η if

$$\|\theta(a)\| \leq \|\eta(a)\|$$

for all $a \in \mathcal{B}$. A derivation component (E,θ) of η is called *maximal* if

$$\|\theta'(a)\| \leq \|\theta(a)\|$$

holds for all derivation components (E',θ') of η.

5.1.6. PROPOSITION. *Two maximal derivation components* $(E^{(i)}, \theta^{(i)})$, $i = 1, 2$, *of* η *are equivalent, i.e. there is a unitary operator*

$$\mathcal{U} : F^{(1)} \to F^{(2)},$$

$F^{(i)}$ *the completion of* $E^{(i)}$, *such that*

$$\mathcal{U}E^{(1)} = E^{(2)}$$

and

$$\theta^{(2)} = \mathcal{U} \circ \theta^{(1)}.$$

PROOF: Since $\|\theta^{(1)}(a)\| = \|\theta^{(2)}(a)\|$ for all $a \in \mathcal{B}$, it follows that $\mathcal{U}\theta^{(1)}(a) = \theta^{(2)}(a)$ gives the desired $\mathcal{U}$. $\diamondsuit$

5.1.7. THEOREM. *The pair* (D_1, η_1) *is a maximal derivation component of* η.

PROOF: We already know that $\eta_1 \in Z^1(D_1, \epsilon)$. We have

$$\|\eta_1(a)\| = \|P_1\eta(a)\| \leq \|\eta(a)\|$$

which shows that η_1 is a derivation component of η.- Proof of maximality: Suppose that (E, θ) is a derivation component of η. Since $\|\theta(a)\| \leq \|\eta(a)\|$ for all $a \in \mathcal{B}$ the Cauchy-Schwartz inequality yields

$$|\langle\theta(a), \theta(b)\rangle| \leq \|\theta(a)\|\|\theta(b)\| \leq \|\eta(a)\|\|\eta(b)\|.$$

Thus

$$(\eta(a), \eta(b))_\theta = \langle\theta(a), \theta(b)\rangle$$

defines a hermitian form on D with $|(u, v)_\theta| \leq \|u\|\|v\|$ for all $u, v \in D$. Therefore, there is a linear operator C on H, $0 \leq C \leq \mathrm{id}$, such that $(u, v)_\theta = \langle Cu, v\rangle$. We have for $a, b \in \mathcal{B}$, $b \in K^2$

$$\langle C\eta(a), \eta(b)\rangle = \langle\theta(a), \theta(b)\rangle = 0,$$

because $\theta \in Z^1(E, \epsilon)$. But this means $C\eta(a) \in H_1$ for all $a \in \mathcal{B}$, so that $C = P_1C$, and, since C is selfadjoint, $P_1CP_1 = C$. Finally, we have for $a \in \mathcal{B}$

$$\begin{aligned}\|\theta(a)\|^2 &= \langle C\eta(a), \eta(a)\rangle \\ &= \langle CP_1\eta(a), P_1\eta(a)\rangle \\ &\leq \|\eta_1(a)\|^2. \diamondsuit\end{aligned}$$

We say that a quadratic g is a *quadratic component* of the conditionally positive hermitian linear functional ψ if $\psi - \mathrm{g}$ is conditionally positive. Then we can write ψ as a sum of a quadratic linear functional and another conditionally positive hermitian linear functional. We say that a quadratic component g of ψ is maximal if $\mathrm{g} - \mathrm{g}'$ is conditionally positive for all quadratic components g' of ψ.

5.1.8. PROPOSITION. *The difference of two maximal quadratic components of ψ is a hermitian left invariant derivation.*

PROOF: Let g_1 and g_2 both be maximal quadratic components of ψ. Then

$$(g_1 - g_2)(a^* a) = 0 \text{ for all } a \in K^1.$$

Since g_1 and g_2 are hermitian, $g_1 - g_2$ is hermitian, too. ◇

For a linear mapping η from $\mathcal{B}$ to a pre-Hilbert space D we denote by $\mathcal{L}(\eta)$ the bilinear form on $\mathcal{B}$ with $\mathcal{L}(\eta)(a,b) = \langle \eta(a^*), \eta(b) \rangle$. We say that $\mathcal{B}$ has the property (C) if

(i) $\mathcal{L}((\eta_\psi)_1) \in B_h^2(\mathbb{C}, \epsilon)$ for all conditionally positive hermitian linear functionals ψ.

Other forms of the same condition (C) are

(ii) For all conditionally positive hermitian linear functionals ψ there is a hermitian linear functional g such that $\partial g = -\mathcal{L}((\eta_\psi)_1)$.

(iii) For all conditionally positive hermitian linear functionals ψ there is a hermitian linear functional g such that

$$g(ab) + g(a)\epsilon(b) - \epsilon(a)g(b) = \langle (\eta_\psi)_1(a^*), (\eta_\psi)_1(b) \rangle.$$

A stronger condition than (C) is given by

$$\mathcal{L}(\theta) \in B^2(\mathbb{C}, \epsilon) \text{ for all } \theta \in Z^1(D, \epsilon). \tag{C'}$$

If $\mathcal{B}$ is finitely generated (C') is equivalent to

$$\mathcal{L}(d) \in B^2(\mathbb{C}, \epsilon) \text{ for all left invariant derivations d.}$$

An apparently even stronger condition is

$$\mathcal{L}(\eta) \in B^2(\mathbb{C}, \epsilon) \text{ for all } \eta \in Z^1(D, \rho). \tag{D}$$

5.1.9. THEOREM. *Let ψ be a conditionally positive hermitian linear functional on $\mathcal{B}$. A hermitian linear functional g satisfying*

$$\partial g = -\mathcal{L}((\eta_\psi)_1)$$

is a maximal quadratic component of ψ. It follows that, if $\mathcal{B}$ has property (C), then all ψ have a maximal quadratic component.

PROOF: Let g be hermitian and $\partial g = -L((\eta_\psi)_1)$. Then for $a \in K^1$

$$g(a^* a) = \|\eta_1(a)\|^2 \le \|\eta(a)\|^2 = \psi(a^* a)$$

where we put $\eta_\psi = \eta$. If g' is quadratic with $g'(a^* a) \le \psi(a^* a)$ it follows that

$$g'(a^* a) = \|\eta_{g'}(a)\|^2 \le \|\eta_1(a)\|^2 = g(a^* a). ◇$$

Our next aim is to prove that each *commutative* $*$-algebra satisfies condition (C). First we show

5.1.10. LEMMA. *Let $\mathcal{B}$ be a $\mathbb{Z}_2$-graded $*$-algebra and let ϵ be even. Let ρ be an even $*$-representation of $\mathcal{B}$ on the $\mathbb{Z}_2$-graded pre-Hilbert space D and let η be in $Z^1(D,\rho)$ and even. Then the following are equivalent*

(i)

$$\langle \eta(a), \eta(b) \rangle = (-1)^{\mathrm{d}_a \mathrm{d}_b} \langle \eta(b^*), \eta(a^*) \rangle$$

for all homogeneous $a, b \in \mathcal{B}$ with $b \in K^2$

(ii)

$$\langle \eta_2(a), \eta_2(b) \rangle = (-1)^{\mathrm{d}_a \mathrm{d}_b} \langle \eta_2(b^*), \eta_2(a^*) \rangle$$

for all homogeneous $a, b \in \mathcal{B}$.

PROOF: (ii) $\Rightarrow$ (i) is clear, because $\eta \lceil K^2 = \eta_2 \lceil K^2$.- (i) $\Rightarrow$ (ii): Let η satisfy (i). For $a, b \in K^2$ we have

$$\langle \eta(a), \eta(b) \rangle = (-1)^{\mathrm{d}_a \mathrm{d}_b} \langle \eta(b^*), \eta(a^*) \rangle.$$

Thus we can define a conjugation $u \mapsto u^*$ on H_2 by

$$\eta(a)^* = \eta(a^*),$$

$a \in K^2$. Moreover, again by (i), we have for $a \in \mathcal{B}$, $b \in K^2$,

$$\begin{aligned}
\langle P_2\eta(a), \eta(b) \rangle &= \langle \eta(a), \eta(b) \rangle \\
&= (-1)^{\mathrm{d}_a \mathrm{d}_b} \langle \eta(b^*), P_2\eta(a^*) \rangle \\
&= \langle (P_2\eta(a^*))^*, \eta(b) \rangle
\end{aligned}$$

which means

$$P_2\eta(a^*) = (P_2\eta(a))^*$$

for all $a \in \mathcal{B}$ and hence for $a, b \in \mathcal{B}$

$$\begin{aligned}
\langle \eta_2(a), \eta(b) \rangle &= \langle P_2\eta(a), P_2\eta(b) \rangle \\
&= \langle (P_2\eta(a^*))^*, P_2\eta(b) \rangle \\
&= (-1)^{\mathrm{d}_a \mathrm{d}_b} \langle P_2\eta(b^*), P_2\eta(a^*) \rangle \\
&= (-1)^{\mathrm{d}_a \mathrm{d}_b} \langle \eta_2(b^*), \eta_2(a^*) \rangle . \diamond
\end{aligned}$$

5.1.11. PROPOSITION. *Let $\mathcal{B}$ be a symmetric $*$-bialgebra. If $\theta \in Z^1(D,\epsilon)$ is such that*

$$\langle \theta(a), \theta(b) \rangle = \langle \theta(b^*), \theta(a^*) \rangle$$

for all $a, b \in \mathcal{B}$, then

$$\mathrm{g} = -\frac{1}{2}(\mathcal{L}(\theta) \circ \Delta)$$

satisfies

$$\partial \mathrm{g} = -\mathcal{L}(\theta).$$

PROOF: Let $a, b \in K^1$. Then

$$\begin{aligned}
&\frac{1}{2}(\mathcal{L}(\theta) \circ \Delta)(ab) \\
&= \frac{1}{2} \sum_{i,j} \langle \theta((b_{1i})^* (a_{1j})^*), \theta(a_{2j} b_{2i}) \rangle \\
&= \frac{1}{2} \sum_{i,j} (\epsilon(b_{1i}) \langle \theta((a_{1j})^*), \theta(a_{2j}) \rangle \epsilon(b_{2i}) + \epsilon(b_{1i}) \langle \theta((a_{1j})^*), \theta(b_{2i}) \rangle \epsilon(a_{2j}) \\
&\quad + \epsilon(a_{1j}) \langle \theta((b_{1i})^*), \theta(b_{2i}) \rangle \epsilon(a_{2j}) + \epsilon(a_{1j}) \langle \theta((b_{1i})^*), \theta(a_{2j}) \rangle \epsilon(b_{2i})) \\
&= \frac{1}{2} (\langle \theta(a^*), \theta(b) \rangle + \langle \theta(b^*), \theta(a) \rangle) \\
&= \langle \theta(a^*), \theta(b) \rangle. \diamond
\end{aligned}$$

5.1.12. THEOREM. *A commutative, symmetric $*$-bialgebra satisfies condition (C).*

PROOF: Let ψ be a conditionally positive hermitian linear functional on the commutative $*$-bialgebra $\mathcal{B}$. For $a, b \in K^1$ we have

$$\psi(ab) = \langle \eta(a^*), \eta(b) \rangle = \psi(ba) = \langle \eta(b^*), \eta(a) \rangle, \tag{5.1.2}$$

and (i) of Lemma 5.1.10 holds. Thus we have (ii) of Lemma 5.1.10. Combining this with (5.1.2) we obtain

$$\langle \eta_1(a), \eta_1(b) \rangle = \eta_1(b^*), \eta_1(a^*) \rangle$$

for all $a, b \in \mathcal{B}$. Now the theorem follows from Prop. 5.1.11.$\diamond$

5.1.12. THEOREM.

(i) *The non-commutative coefficient algebra $\mathcal{U}\langle n \rangle$ has the property (D). For $\eta \in Z^1(D, \rho)$ a hermitian solution ψ of the problem*

$$\partial \psi = -\mathcal{L}(\eta)$$

is determined by the equations

$$\psi(1) = 0$$

$$\psi(x_{kl}) = -\frac{1}{2} \sum_{m=1}^{n} \langle \eta(x^*_{km}), \eta(x^*_{lm}) \rangle.$$

(ii) *For $L \in M_n$, a left invariant derivation on $\mathcal{U}\langle n \rangle$ is determined by the equations*

$$\mathrm{d}_L(x_{kl}) = L_{kl} \tag{5.1.3}$$

$$\mathrm{d}_L(x^*_{kl}) = -L_{lk}. \tag{5.1.4}$$

This establishes a 1-1-correspondence between M_n and $Z^1(\mathbb{C}, \epsilon)$. A left invariant derivation d_L is hermitian if and only if L is skew-hermitian.

(iii) *For $L \in M_n$ a hermitian solution g of the problem*

$$\partial \mathrm{g} = -\mathcal{L}(\mathrm{d}_L)$$

is determined by the equations

$$g(1) = 0$$
$$g(x_{kl}) = -\frac{1}{2}(L^* L)_{kl}.$$

PROOF: (i) follows from Prop. 4.4.7 if we put $R = 0$.

(ii): Define the left invariant derivation $\tilde{d}_L$ on F by $\tilde{d}_L(x_{kl}) = L_{kl}$ and $\tilde{d}_L(x^*_{kl}) = -L_{lk}$. Then

$$\begin{aligned}\tilde{d}_L(\sum_{m=1}^n x_{km}x^*_{lm}) &= \sum_{m=1}^n (L_{km}\delta_{lm} + \delta_{km}(-L_{ml})) \\ &= 0 \\ &= \tilde{d}_L(\sum_{m=1}^n x^*_{mk}x_{ml})\end{aligned}$$

which shows that $\tilde{d}_L$ respects the relations in $\mathcal{U}\langle n\rangle$. Thus it gives rise to a left invariant derivation d_L on $\mathcal{U}\langle n\rangle$ satisfying (5.1.3) and (5.1.4). On the other hand, a left invariant derivation d on $\mathcal{U}\langle n\rangle$ must satisfy $d(x_{kl}) = -d(x^*_{lk})$. (iii): This is just (i) for the special case $\eta = d_L$.◇

There are a lot of conditionally positive hermitian linear functionals with no quadratic component. Let φ be a positive linear funtional on $\mathcal{B}$. Then $\psi = \varphi - \varphi(1)\epsilon$ is conditionally positive hermitian and vanishes at 1. It is not difficult to see that ψ contains no quadratic component, and ψ might be said to be of 'Poisson type'; cf. [27]. However, the generator $g_{\kappa,\beta}$ of the non-commutative Poisson process of Section 4.3 is not of this type. Since $\beta_i \neq 0$ for at least one i, it follows that $g_{\kappa,\beta}$ has no quadratic component.

5.2. Infinitely divisible states on the Weyl algebra. Let V be a $\mathbb{Z}_2$-graded involutive vector space. In this section, all linear functionals are understood to be even. The left invariant derivations on $\mathcal{T}(V)$ can be identified with the linear functionals β on V via $d_\beta \lceil V = \beta$. A left invariant derivation d_β is hermitian if and only if β is hermitian. The quadratic linear functionals on $\mathcal{T}(V)$ vanishing on V can be identified with the positive sesquilinear forms Q on V via $g_Q(vw) = Q(v^*, w)$ for $v, w \in V$. Since there are no relations in $\mathcal{T}(V)$ it is clear that $\mathcal{T}(V)$ satisfies condition (D). For, if η is in $Z^1(D, \rho)$ a solution of $\partial\psi = -\mathcal{L}(\eta)$ is simply given by

$$\begin{aligned}\psi(1) &= 0 \\ \psi(v) &= 0 \\ \psi(v_1 \ldots v_n) &= \langle \eta(v_1^*), \eta(v_2 \ldots v_n)\rangle,\end{aligned}$$

$v, v_1 \ldots, v_n \in V$, $n \geq 2$. It follows that all conditionally positive hermitian linear functionals on $\mathcal{T}(V)$ have maximal quadratic components. This result was already obtained in [5]. Let C be a bilinear form on V satisfying

$$C(v, w) = -(-1)^{d_v d_w} C(w, v) = \overline{C(w^*, v^*)}$$

We denote by $I(C)$ the ideal in $T(V)$ generated by the elements $[v,w]-C(v,w)$, $v,w\in V$. This makes sense, because $C(v,w)=S([v,w])$ with $S:V\otimes V\to\mathbb{C}$ the linear mapping given by $S(v\otimes w)=\frac{1}{2}C(v,w)$. Moreover, $I(C)$ is a $*$-ideal. The $*$-algebra $T(V)/I(C)$ is called the *Weyl algebra* $\mathrm{W}_C=\mathrm{W}_C(V)$ associated to C. If $C\neq 0$ the coalgebra structure of $T(V)$ cannot be transferred to W_C because $I(C)$ is not a coideal. A state φ on W_C is called infinitely divisible if the state $\tilde{\varphi}$ on $T(V)$ induced by φ via the canonical mapping from $T(V)$ to W_C is infinitely divisible. In other words, the infinitely divisible states on W_C are those infinitely divisible states on $T(V)$ that vanish on $I(C)$. We show in this section that an infinitely divisible state φ on W_C must be of the form $\varphi=\mathrm{g}_Q\star\mu$ where μ vanishes on $\mathrm{W}_0(V)$ and Q is such that

$$Q(v^*,w)-(-1)^{\mathrm{d}_v\mathrm{d}_w}Q(w^*,v)=C(v,w).$$

This is the infinitesimal, $\mathbb{Z}_2$-graded version of a result of H. Araki [8] for the Heisenberg group. If the grading of V is trivial $\mu\lceil\mathrm{W}_0(V)=0$ means that μ is 'classical' because $\mathrm{W}_0(V)$ is the ideal that makes $T(V)$ commutative. If $V_{(0)}=\{0\}$ it means $\mu=0$.

We begin with a simple but useful proposition; see also [33, 56, 86].

5.2.1.PROPOSITION. *For a normalized linear functional φ on $T(V)$ and $n\in\mathbb{N}$ there is a uniquely determined normalized linear functional φ_n on $T(V)$ such that $\varphi=(\varphi_n)^{\star n}$.*

PROOF: We define φ_n by the recursion formula

$$n\varphi_n(v_1\dots v_m)=\varphi(v_1\dots v_m)-\sum_{A_1,\dots A_n}\operatorname{sgn}\varphi_n(v_{A_1})\dots\varphi_n(v_{A_n}) \tag{5.2.1}$$

where the sum is extended over all ordered disjoint partitions $(A_1,\dots,A_n)$ of the set $\{1,\dots,m\}$ with $A_j\neq\{1,\dots,m\}$ (but $A_j=\emptyset$ is allowed). For $A=\{j_1<\dots<j_k\}\subset\{1,\dots,m\}$ the symbol v_A denotes the monomial $v_{j_1}\dots v_{j_k}$ (we put $v_\emptyset=1$!). The sign appearing under the sum depends on the partition and on the degrees of the v_i. It is equal to

$$\operatorname{sgn}(\pi;\mathrm{d}(v_1),\dots,\mathrm{d}(v_m))$$

with π the permutation

$$\pi=\begin{pmatrix}1&\dots&m\\A_1&\dots&A_n\end{pmatrix}$$

where in the lower row we write the sets A_k in their natural order; see Section 4.5 for the definition of sgn(;). It is clear that $(\varphi_n)^{\star n}=\varphi$ and that, on the other hand, a normalized φ_n with $(\varphi_n)^{\star n}=\varphi$ has to satisfy (5.2.1).$\diamond$

If the linear functional ψ on $T(V)$ vanishes at 1 we have $\psi^{\star n}(M)=0$ for all monomials of length less than n, Therefore, the series

$$\exp_\star\psi=\sum_{n=0}^{\infty}\frac{\psi^{\star n}}{n!}$$

can be defined pointwise as a normalized linear functional on $\mathcal{T}(V)$. For the same reasons,

$$\ln_\star \varphi = \sum_{n=0}^{\infty} (-1)^{n+1} \frac{(\varphi - \epsilon)^{\star n}}{n}$$

makes sense for a normalized φ and is a linear functional vanishing at $\mathbf{1}$. Moreover, $\ln_\star$ is the inverse of $\exp_\star$.

5.2.2. THEOREM. *A state φ on $\mathcal{T}(V)$ is infinitely divisible if and only if there is a conditionally positive, hermitian linear functional ψ on $\mathcal{T}(V)$ such that $\varphi = \exp_\star \psi$.*

PROOF: Let φ be infinitely divisible. We must prove that $\ln_\star \varphi$ is conditionally positive and hermitian. But we have

$$(\exp_\star(\frac{1}{n} \ln_\star \varphi))^{\star n} = \varphi$$

which by Prop. 5.2.1 yields

$$\exp_\star(\frac{1}{n} \ln_\star \varphi) = \varphi_n$$

and $n(\varphi_n - \epsilon)$ converges to $\ln_\star \varphi$ pointwise. Since $n(\varphi_n - \epsilon)$ all are conditionally positive and hermitian, the same holds for $\ln_\star \varphi$.- The converse follows from Schoenberg correspondence.$\diamond$

Next we find necessary and sufficient conditions on ψ for $\exp_\star \psi$ to vanish on $I(C)$. We need the following lemma.

5.2.3. LEMMA. *Let $\varphi, \varphi_1, \dots, \varphi_n$ be normalized linear functionals on $\mathcal{T}(V)$.*
(i) *If φ_j vanishes on $I(C_j)$ then $\varphi_1 \star \cdots \star \varphi_n$ vanishes on $I(C_1 + \cdots + C_n)$.*
(ii) *If $\varphi = (\varphi_n)^{\star n}$ and φ vanishes on $I(C)$ then φ_n vanishes on $I(\frac{1}{n}C)$.*

PROOF: (i): We have for $a, b \in \mathcal{T}(V)$ and $v, w \in V$

$$\begin{aligned}
&(\varphi_1 \star \cdots \star \varphi_n)(a([v,w] - \sum_{j=1}^{n} C_j(v,w))b) \\
&= (\varphi_1 \otimes \ldots \otimes \varphi_n)(\Delta_n(a)(([v,w] - C_1(v,w)) \otimes \mathbf{1} \otimes \cdots \otimes \mathbf{1} \\
&\quad + \mathbf{1} \otimes (([v,w] - C_2(v,w)) \otimes \mathbf{1} \otimes \cdots \otimes \mathbf{1} + \ldots \\
&\quad \cdots + \mathbf{1} \otimes \cdots \otimes \mathbf{1} \otimes ([v,w] - C_n(v,w)))\Delta_n(b)) \\
&= \sum_{i,j} (\varphi_1(a_{1i}([v,w] - C_1(v,w))b_{1j})\varphi_2(a_{2i}b_{2j}) \ldots \varphi_n(a_{ni}b_{nj}) + \ldots \\
&\quad \cdots + \varphi_1(a_{1i}b_{1j}) \ldots \varphi_{n-1}(a_{(n-1),i}b_{(n-1),j})\varphi_n(a_{ni}([v,w] - C_n(v,w))b_{nj}) \\
&= 0
\end{aligned}$$

if φ_j vanishes on $I(C_j)$. (ii): Suppose φ vanishes on $I(C)$ and

$$\varphi_n(M([v,w] - \frac{1}{n}C(v,w))N) = 0$$

for all $v, w \in V$ and all monomials M, N with the length of MN less than m. Then for monomials M, N with the length of MN equal to m we have

$$\begin{aligned} 0 &= \varphi(M([v,w] - C(v,w))N) \\ &= (\varphi_n)^{\star n}(M([v,w] - C(v,w))N) \\ &= n\varphi_n(M([v,w] - \frac{1}{n}C(v,w))N). \diamond \end{aligned}$$

5.2.4. THEOREM. *Let φ be an infinitely divisible state on $\mathcal{T}(V)$. Then the following are equivalent*

(i) *φ vanishes on $I(C)$.*

(ii)

$$\ln_\star \varphi([v,w]) = C(v,w) \tag{5.2.2}$$

and

$$\ln_\star \varphi(a[v,w]b) = 0 \tag{5.2.3}$$

for all $v, w \in V$ and $a, b \in \mathcal{T}(V)$ such that $ab \in K^1$.

PROOF: (i) $\Rightarrow$ (ii): We have

$$\ln_\star \varphi([v,w]) = \lim_{n\to\infty} n\varphi_n([v,w]) = C(v,w).$$

For $a, b \in \mathcal{T}(V)$ with $ab \in K^1$

$$\begin{aligned} \ln_\star \varphi(a[v,w]b) &= \lim_{n\to\infty} n\varphi_n(a[v,w]b) \\ &= \lim_{n\to\infty} \varphi_n(ab)C(v,w) \\ &= 0 \end{aligned}$$

because

$$\begin{aligned} \lim_{n\to\infty} \varphi_n(a) &= \lim_{n\to\infty} \frac{1}{n}(n\varphi_n - \epsilon)(a) \\ &= \lim_{n\to\infty} \frac{1}{n}\psi(a) \\ &= 0 \end{aligned}$$

for $a \in K^1$.- (ii) $\Rightarrow$ (i): For $a, b \in \mathcal{T}(V)$ and $v, w \in V$ we have

$$\begin{aligned} &(\ln_\star \varphi)^{\star(n+1)}(a([v,w] - C(v,w))b) \\ &= (\ln_\star \varphi)^{\otimes(n+1)}(\Delta_{n+1}(a)([v,w] \otimes \mathbf{1} \otimes \cdots \otimes \mathbf{1} + \ldots \\ &\qquad \cdots + \mathbf{1} \otimes \cdots \otimes \mathbf{1} \otimes [v,w])\Delta_{n+1}(b)) \\ &\qquad - C(v,w)(\ln_\star \varphi)^{\star(n+1)}(ab) \\ &= (n+1)C(v,w)(\ln_\star \varphi)^{\star n}(ab) - C(v,w)(\ln_\star \varphi)^{\star(n+1)}(ab) \end{aligned}$$

so that

$$\begin{aligned}
&\varphi(a([v,w]-C(v,w))b)\\
&=\sum_{n=0}^{\infty}\frac{(\ln_\star\varphi)^{\star n}}{n!}(a([v,w]-C(v,w))b)\\
&=-\epsilon(a)\epsilon(b)C(v,w)+\sum_{n=0}^{\infty}\frac{(\ln_\star\varphi)^{\star(n+1)}}{(n+1)!}(a([v,w]-C(v,w))b)\\
&=-\epsilon(a)\epsilon(b)C(v,w)+\Big(\sum_{n=0}^{\infty}\frac{(n+1)(\ln_\star\varphi)^{\star n}}{(n+1)!}(ab)\\
&\qquad-\sum_{n=0}^{\infty}\frac{(\ln_\star\varphi)^{\star(n+1)}}{(n+1)!}(ab)\Big)C(v,w)\\
&=0.\diamondsuit
\end{aligned}$$

We say that a quadratic g on $\mathcal{T}(V)$ is *centralized* if it vanishes on V. Clearly, each generator on $\mathcal{T}(V)$ has a unique centralized maximal quadratic component.

5.2.5. THEOREM. *Let ψ be a generator on $\mathcal{T}(V)$ satisfying the conditions (5.2.2) and (5.2.3) of Theorem 5.2.4 (ii). Let*

$$\psi = g_Q + \mu$$

be the unique decomposition of ψ into a centralized maximal quadratic component g_Q and a conditionally positive, hermitian linear functional μ with degenerate maximal quadratic component. Then

(i) $Q(v^*,w)-(-1)^{d_v d_w}Q(w^*,v)=C(v,w)$.

(ii) *μ vanishes on $I(0)$.*

PROOF: It follows from (5.2.3) that

$$\psi(abc)=(-1)^{(d_a+d_b)d_c}\psi(cab)$$

for $a,b,c\in K^1$. But this means

$$\langle\eta(ab),\eta(c)\rangle=(-1)^{(d_a+d_b)d_c}\langle\eta(c^*),\eta(b^*a^*)\rangle$$

for $a,b\in K^1$, $c\in\mathcal{T}(V)$. By Lemma 5.1.10 this implies

$$\langle\eta_2(a),\eta_2(b)\rangle=(-1)^{d_a d_b}\langle\eta_2(b^*),\eta_2(a^*)\rangle$$

for $a,b\in\mathcal{T}(V)$, or

$$\mu\,(ab)=\mu(ba) \tag{5.2.4}$$

for $a,b\in K^1$. Moreover, μ agrees with ψ on elements of the form $a[v,w]b$ with $ab\in K^1$, $v,w\in V$. Together with (5.2.4) this implies (ii). By (5.2.2) we have

$$\psi([v,w])=C(v,w)=g_Q([v,w])=Q(v^*,w)-(-1)^{d_v d_w}Q(w^*,v)$$

which is (i).$\diamondsuit$

5.2.6. COROLLARY. *The infinitely divisible states on* $W_C(V)$ *are precisely the states on* $T(V)$ *of the form*

$$\gamma_Q \star \mu$$

where

$$Q(v^*, w) - (-1)^{d_v d_w} Q(w^*, v) = C(v, w) \tag{5.2.5}$$

and μ *is an infinitely divisible state on* $W_0(V)$ *which cannot be decomposed into a convolution product of a non-degenerate gaussian state and another infinitely divisible state on* $W_0(V)$.◇

If $C \neq 0$ it follows from relation (5.2.5) for the covariance matrix of the maximal gaussian component that an infinitely divisible state on $W_C(V)$ has a non-degenerate maximal gaussian component.

If $V_{(0)} = V$ we have that μ lives on the commutative tensor algebra over V, and the only non-commutative part of an infinitely divisible state is its maximal gaussian component.

If $V_{(0)} = \{0\}$, i.e. $d_v = 1$ for all $v \in V$, we have $\mu(v^2) = 0$ for all $v \in V$. Using the fact that any $v \in V$ can be written as the sum of hermitian elements and the Cauchy-Schwartz inequality, this implies $\mu = 0$. Thus our result includes the result of D. Mathon and R.L. Streater [56] that an infinitely divisible state on the algebra of the canonical anti-commutation relations must be gaussian.

6. Limit theorems

Using a convergence principle for convolution products of linear functionals on coalgebras, we prove that the correlations of the underlying additive white noise of a given white noise can be approximated by sums of normalized increments. This is the quantum analogue of a well-known result for matrix-valued classical stochastic processes; see e.g. [81] and the introduction of these notes. The same convergence principle gives a central limit theorem, or more precisely an algebraic analogue of the Donsker Brownian motion invariance principle, for sums of identically distributed independent quantum random vectors. The limiting processes are gaussian white noise in the sense of Chapter 5 and, in the case of q-additive white noise, the limits are Azéma noise or q-deformations of quantum Brownian motion which were introduced in Chapter 4.

6.1. A coalgebra limit theorem. The following simple result will be useful.

6.1.1. THEOREM. *Let φ_{nk}, $n \in \mathbb{N}$, $k = 1, \ldots, k_n$ ($k_n \in \mathbb{N}$), be linear functionals on a coalgebra $\mathcal{C}$ satisfying*

(i) *$\varphi_{n1}, \ldots, \varphi_{nk_n}$ commute for each $n \in \mathbb{N}$ with respect to convolution*

(ii)

$$\lim_{n\to\infty} \max_{1\leq k\leq k_n} |(\varphi_{nk} - \epsilon)(c)| = 0$$

for all $c \in \mathcal{C}$.

(iii)

$$\sup_{n\in\mathbb{N}} \sum_{1\leq k\leq k_n} |(\varphi_{nk} - \epsilon)(c)| < \infty$$

for all $c \in \mathcal{C}$.

Furthermore, suppose that there is a linear functional ψ on $\mathcal{C}$ such that for all $c \in \mathcal{C}$

$$\lim_{n\to\infty} \Big(\sum_{1\leq k\leq k_n} (\varphi_{nk} - \epsilon)(c)\Big) = \psi(c). \tag{6.1.1}$$

Then

$$\lim_{n\to\infty} \Big(\prod_{1\leq k\leq k_n}^{*} \varphi_{nk}\Big)(c) = (\exp_\star \psi)(c) \tag{6.1.2}$$

for all $c \in \mathcal{C}$ (where the product $\prod^*$ is the convolution product).

PROOF: Let (φ_{nk}) satisfy (i), (ii), (iii) and (6.1.1). For $c \in \mathcal{C}$ there is a finite-dimensional sub-coalgebra $\mathcal{D}_c$ of $\mathcal{C}$ containing c by the fundamental theorem on coalgebras. The linear operators $T_{nk} = (\mathrm{id} \otimes \varphi_{nk}) \circ \Delta$ on $\mathcal{C}$ leave $\mathcal{D}_c$ invariant. By (ii) and (iii)

$$\lim_{n\to\infty} \max_{1\leq k\leq k_n} \|T_{nk} \lceil \mathcal{D}_c - \mathrm{id}\| = 0$$

and

$$\sup_{n\in\mathbb{N}} \sum_{1\leq k\leq k_n} \|T_{nk} \lceil \mathcal{D}_c - \mathrm{id}\| < \infty.$$

Moreover, by (6.1.1)

$$\lim_{n\to\infty} \sum_{1\le k\le k_n} (T_{nk}\lceil \mathcal{D}_c - \mathrm{id}) = R\lceil \mathcal{D}_c$$

with $R = (\mathrm{id}\otimes\psi)\circ\Delta$. Since by (i) the operators $T_{n1},\ldots,T_{nk_n}$ commute for each $n\in\mathbb{N}$ we can use the same arguments as for complex numbers (see e.g. [21] p. 184) to obtain

$$\lim_{n\to\infty} \prod_{1\le k\le k_n} T_{nk}\lceil \mathcal{D}_c = \exp(R\lceil \mathcal{D}_c). \tag{6.1.3}$$

If we apply the counit ϵ to both sides of (6.1.3) and evaluate at c this gives (6.1.2).◇

As a consequence we have the following

6.1.2. **Theorem**. *Let φ_n, $n\in\mathbb{N}$, be linear functionals on a coalgebra $\mathcal{C}$ such that*

$$\lim_{n\to\infty} n(\varphi_n - \epsilon)(c) = \psi(c)$$

for some linear functional ψ on $\mathcal{C}$ and for all $c\in\mathcal{C}$. Then

$$\lim_{n\to\infty} (\varphi_n)^{\star n}(c) = (\exp_\star \psi)(c)$$

for all $c\in\mathcal{C}$.

Proof: This is immediate from Theorem 6.1.1 if we put $k_n = n$ and $\varphi_{nk} = \varphi_n$.◇

For an $\mathbb{N}$-graded coalgebra $\mathcal{C}$ and a complex number z we define the linear operator $s(z)$ on $\mathcal{C}$ by

$$s(z)c = z^{\mathrm{l}(c)}c$$

where $\mathrm{l}(c)$ denotes the degree of c. Then for linear functionals φ and ψ on $\mathcal{C}$

$$(\varphi\star\psi)\circ s(z) = (\varphi\circ s(z))\star(\psi\circ s(z)).$$

6.1.3. **Theorem**. *Let $\mathcal{C}$ be an $\mathbb{N}$-graded coalgebra and let $\kappa\in\mathbb{N}$. If a linear functional φ on $\mathcal{C}$ satisfies*

(i) $\varphi\lceil\mathcal{C}^{(l)} = 0$ *for* $0 < l < \kappa$

(ii) $\varphi\lceil\mathcal{C}^{(0)} = \epsilon\lceil\mathcal{C}^{(0)}$

then for all $c\in\mathcal{C}$

$$\lim_{n\to\infty} (\varphi^{\star n}\circ s(n^{-\frac{1}{\kappa}}))(c) = (\exp_\star g_\varphi)(c),$$

where g_φ denotes the linear functional on $\mathcal{C}$ with

$$g_\varphi\lceil\mathcal{C}^{(l)} = 0 \text{ for all } l\neq k$$
$$g_\varphi\lceil\mathcal{C}^{(\kappa)} = \varphi\lceil\mathcal{C}^{(\kappa)}.$$

Proof: This is a corollary to Theorem 6.1.2. For, if we put $\varphi_n = \varphi\circ s(n^{-\frac{1}{\kappa}})$ then for $c\in\mathcal{C}$ with $\mathrm{l}(c)\ge\kappa$

$$n(\varphi_n - \epsilon)(c) = n\varphi_n(c) = n\cdot n^{-\frac{\mathrm{l}(c)}{\kappa}}\varphi(c)$$

which is equal to $\varphi(c)$ if $\mathrm{l}(c)=\kappa$ and tends to 0 for $\mathrm{l}(c)>\kappa$. Moreover, by assumption $n(\varphi_n-\epsilon)(c) = 0$ for $\mathrm{l}(c)<\kappa$.◇

For $\kappa = 1$ the above theorem is a law of large numbers and for $\kappa = 2$ it is a central limit theorem; cf. [28, 90] where the special case of a graded tensor algebra with 'Boson and Fermion convolution' was treated; cf. also [2, 33, 38].

6.2. The underlying additive noise as a limit. By a partition of $\mathbb{R}_+$ we mean a monotone sequence (t_l) in $\mathbb{R}_+$ with $t_0 = 0$ and $t_l \to \infty$ for $l \to \infty$. We put $\Delta t_l = t_l - t_{l-1}$.

6.2.1. THEOREM. *Let $(j_{st})_{(s,t)\in T_L}$ be the realization on Bose Fock space of a white noise on the involutive (γ, α)-bialgebra $\mathcal{B}$. For a partition $\mathcal{Z}$ of $\mathbb{R}_+$ and for $b \in \mathcal{B}$ we put*

$$S_{Z,t}(b) = \sum_{0<t_l\leq t} (j_{t_{l-1}t_l}(b) - \epsilon(b)\mathrm{id}).$$

Then for a series $\mathcal{Z}_{(n)}$ of partitions of $\mathbb{R}_+$ with

$$\lim_{n\to\infty} \max_{l\in\mathbb{N}} \Delta t_l^{(n)} = 0$$

the additive processes $S_{n,t} = S_{\mathcal{Z}^{(n)},t}$ converge strongly on $\mathcal{E}_D$ to the additive white noise F_t given by j_{st} for $n \to \infty$.

PROOF: Denote by $s_{n,t}(b)$ the kernel

$$\sum_{0<t_l^{(n)}\leq t} (k_{t_{l-1}^{(n)} t_l^{(n)}}(b) - \epsilon(b)\Omega^{(3)})$$

where $k_{st} = k_{st}^{(\gamma,\alpha)} \star \gamma_{t,\infty}(\rho(\alpha_{\mathrm{d}(b)}))$, and denote by $f_t(b)$ the kernel of $F_t(b)$ as given by the basic formula. Let $\sigma, \tau, \varrho \in \mathcal{S}$ be disjoint and $\sigma \cup \tau \cup \varrho \subset [s, t)$. Then for n sufficiently large

$$\begin{aligned} &s_{n,t}(b)(\sigma, \tau, \varrho) \\ &\quad = (\varphi_{r-t_n} \star \kappa(r; \sigma, \tau, \varrho) \star \varphi_{t_{n+1}-r})(b) \prod_{k=1}^{n} \lambda(r_k)(\rho(\alpha_{\mathrm{d}(b)}) - \mathrm{id}) \end{aligned}$$

if $\sigma = \{r\}$, $t_n \leq r < t_{n+1}$, and $\tau = \{r_1 < \cdots < r_n\}$ with $r < r_1$, and $s_{n,t}(b)(\sigma, \tau, \varrho) = 0$ otherwise. Thus

$$\lim_{n\to\infty} s_{n,t}(b)(\sigma, \tau, \varrho) = f_t(b)(\sigma, \tau, \varrho).$$

Moreover,

$$s_{n,t}(b)(\emptyset, \emptyset, \emptyset) = \sum_{0<t_l^{(n)}\leq t} (\varphi_{t_l^{(n)} - t_{l-1}^{(n)}}(b) - \epsilon(b))$$

which converges to $t\psi(b)$ for $n \to \infty$. Using an argument similar to that of the proof of Prop. 2.3.1, one shows that $s_{n,t}(b)$ also satisfy the other conditions of Prop. 2.2.2.◇

Using the results of the previous section, we prove the *convergence of the correlations* which is the convergence of interest in the frame-work of algebraic quantum probability theory.

Let ψ be a generator on the $*$-bialgebra $\mathcal{B}$ and let (D, η, ρ, ψ) be the quadruple associated with ψ. If we put $\eta_{\mathrm{a},0} = \eta$, $\rho_{\mathrm{a},0} = \rho - \epsilon\,\mathrm{id}$ and $\psi_{\mathrm{a},0} = \psi$, then $(D, \eta_{\mathrm{a},0}, \rho_{\mathrm{a},0}, \psi_{\mathrm{a},0})$

gives rise to a generator ψ_{a} on the involutive tensor (γ,α)-Hopf algebra $\mathcal{T}(\mathcal{B})$ over the double-module $*$-vector space $\mathcal{B}$; see Section 4.2.

6.2.2. PROPOSITION. *For $l \in \mathbb{N}$ and $b_1,\ldots,b_l \in \mathcal{B}$ the equality*

$$\psi_{\mathrm{a}}(b_1\ldots b_l) = \psi((b_1 - \epsilon(b_1)\mathbf{1}_{\mathcal{B}})\ldots(b_l - \epsilon(b_l)\mathbf{1}_{\mathcal{B}}))$$

holds.

PROOF: First

$$\psi_{\mathrm{a}}(b) = \psi_{\mathrm{a},0}(b) = \psi(b) = \psi(b - \epsilon(b)\mathbf{1}_{\mathcal{B}})$$

because of $\psi(\mathbf{1}_{\mathcal{B}}) = 0$. Next for $l \geq 2$ and $b_1,\ldots,b_l \in \mathcal{B}$:

$$\begin{aligned}
&\psi_{\mathrm{a}}(b_1\ldots b_l)\\
&= \langle \eta_{\mathrm{a}}(b_1^*), \eta_{\mathrm{a}}(b_2\ldots b_l)\rangle\\
&= \langle \eta(b_1^*), \rho_{\mathrm{a}}(b_2\ldots b_{l-1})\eta(b_l)\rangle\\
&= \langle \eta(b_1^*), \rho(b_2 - \epsilon(b_2)\mathbf{1}_{\mathcal{B}})\ldots\rho(b_{l-1} - \epsilon(b_{l-1})\mathbf{1}_{\mathcal{B}})\eta(b_l - \epsilon(b_l)\mathbf{1}_{\mathcal{B}})\rangle\\
&= \langle \eta(b_1^*), \eta((b_2 - \epsilon(b_2)\mathbf{1}_{\mathcal{B}})\ldots(b_l - \epsilon(b_l)\mathbf{1}_{\mathcal{B}}))\rangle\\
&= \psi(b_1(b_2 - \epsilon(b_2)\mathbf{1}_{\mathcal{B}})\ldots(b_l - \epsilon(b_l)\mathbf{1}_{\mathcal{B}}))\\
&\qquad - \epsilon(b_1)\psi((b_2 - \epsilon(b_2)\mathbf{1}_{\mathcal{B}})\ldots(b_l - \epsilon(b_l)\mathbf{1}_{\mathcal{B}}))
\end{aligned}$$

where we used $\eta(\mathbf{1}_{\mathcal{B}}) = 0$ and the facts that η_{a} is a 1-cocycle with respect to the representation ρ_{a} and η is a 1-cocycle with respect to the representation ρ.◇

In order to simplify things (we want condition (i) of Theorem 6.1.1 to be satisfied), we only consider equidistant partitions. For $t \in \mathbb{R}$ the Gauss symbol $[t]$ means the biggest integer smaller than t.

6.2.3. THEOREM. *The correlations of the additive process $S_{n,t}$ given by*

$$S_{n,t}(b) = \sum_{l=1}^{[nt]}(j_{(l-1)/n,l/n}(b) - \epsilon(b)\mathrm{id})$$

converge to the correlations of the additive white noise F_t associated with j_{st}. In other words,

$$\lim_{n\to\infty} \Phi(S_{n,t_1}(b_1)\ldots S_{n,t_m}(b_m)) = \Phi(F_{t_1}(b_1)\ldots F_{t_m}(b_m))$$

for all $m \in \mathbb{N}$, $t_1,\ldots,t_m \in \mathbb{R}_+$ and $b_1,\ldots,b_m \in \mathcal{B}$.

PROOF: The additive increments $S_{n,st} = S_{n,t} - S_{n,s}$ of $S_{n,t}$ are independent because $b - \epsilon(b)\mathbf{1}$ is homogeneous if b is homogeneous and in this case has the same degree as b, and because

$$S_{n,st}(b) = \sum_{l=[ns]+1}^{[nt]} j_{(l-1)/n,l/n}(b - \epsilon(b)\mathbf{1}).$$

This means that it is sufficient to prove

$$\lim_{n\to\infty} \Phi(S_{n,t}(b_1)\ldots S_{n,t}(b_m)) = \Phi(F_t(b_1)\ldots F_t(b_m))$$

for all $m \in \mathbb{N}$, $t \in \mathbb{R}_+$ and $b_1, \ldots, b_m \in \mathcal{B}$. The quantum random vector $S_{n,t}$ on the double-module $*$-vector space $\mathcal{B}$ is the sum of the independent tuple

$$(j_{0,1/n} - \epsilon\mathrm{id}, \ldots, j_{([nt]-1)/n,[nt]/n} - \epsilon\mathrm{id})$$

of quantum random vectors on $\mathcal{B}$, and therefore

$$\Phi(S_{n,t}(b_1) \ldots S_{n,t}(b_m)) = (\varphi_n)^{\star_a [nt]}(b_1 \ldots b_m)$$

where φ_n is the distribution on $\mathcal{T}(\mathcal{B})$ of the quantum random vector $j_{0,1/n} - \epsilon\mathrm{id}$ and $\star_a$ denotes the convolution of the involutive tensor (γ, α)-Hopf algebra $\mathcal{T}(\mathcal{B})$. We have

$$\begin{aligned} \varphi_n(b_1 \ldots b_m) &= \Phi((j_{0,1/n}(b_1) - \epsilon\mathrm{id}) \ldots (j_{0,1/n}(b_m) - \epsilon\mathrm{id})) \\ &= \Phi(j_{0,1/n}(b_1 - \epsilon(b_1)\mathbf{1}) \ldots (b_m - \epsilon(b_m)\mathbf{1}))) \\ &= \varphi_{1/n}((b_1 - \epsilon(b_1)\mathbf{1}) \ldots (b_m - \epsilon(b_m)\mathbf{1})). \end{aligned}$$

Moreover, for $b \in \mathrm{Kern}\ \epsilon$

$$\lim_{n \to \infty} [nt]\varphi_{1/n}(b) = t\psi(b),$$

so that

$$\begin{aligned} \lim_{n \to \infty} [nt]\varphi_n(b_1 \ldots b_m) &= \lim_{n \to \infty} [nt]\varphi_{1/n}((b_1 - \epsilon(b_1)\mathbf{1}) \ldots (b_m - \epsilon(b_m)\mathbf{1})) \\ &= t\psi((b_1 - \epsilon(b_1)\mathbf{1}) \ldots (b_m - \epsilon(b_m)\mathbf{1})) \\ &= t\psi_a(b_1 \ldots b_m). \end{aligned}$$

Now it follows from Theorem 6.1.2 that

$$\begin{aligned} \lim_{n \to \infty} (\varphi_n)^{\star_a [nt]}(b_1 \ldots b_m) &= (\exp_{\star_a} t\psi_a)(b_1 \ldots b_m) \\ &= \Phi(F_t(b_1) \ldots F_t(b_m)) \end{aligned}$$

which completes the proof.◇

6.3. Invariance principles. We apply the results of Section 6.1 to tensor Hopf algebras and their symmetrizations.

As in Chapter 4 let $\mathbb{L}$ be a $*$-group, V a double-module $*$-vector space, i.e. V is an $\mathbb{L}$-graded $*$-vector space which is also a $\mathbb{C}\mathbb{L}$-module in such a way that

$$\mathrm{d}(\lambda.v) = \lambda \mathrm{d}(v)\lambda^{-1} \text{ and } (\lambda.v)^* = v^*.\lambda^*$$

for $\lambda \in \mathbb{L}$, $v \in V$ homogeneous. Clearly, the involutive (γ, α)-Hopf algebra $\mathcal{T}(V)$ is also an $\mathbb{N}$-graded coalgebra if we define the degree $\mathrm{l}(b)$ of a homogeneous polynomial to be its usual polynomial degree.

6.3.1. THEOREM. *Let $(F_n)_{n \in \mathbb{N}}$ be a left independent sequence of centralized identically distributed quantum random vectors on V. For $v \in V$ we put*

$$S_{n,t} = \frac{1}{\sqrt{n}} \sum_{l=1}^{[nt]} F_l.$$

Then the correlations of the vector process $S_{n,t}$ on V converge to the correlations of the additive left white noise on V given by the quadratic generator g_Q of Section 4.3 with $Q(v,w) = \varphi(v^* w)$ where φ denotes the distribution of F_l.

PROOF: It is easy to see that the additive increments $(S_{n,st})_{(s,t)\in T_{\mathbb{L}}}$ are independent. One only has to remark that $s \mapsto [s]$ is a monotone function. Thus it remains to prove that

$$\lim_{n\to\infty} \varphi_n^{\star[nt]}(v_1 \ldots v_m) = (\exp_\star t g_Q)(v_1 \ldots v_m)$$

for all $m \in \mathbb{N}$, $v_1, \ldots, v_m \in V$, where $\varphi_n = \varphi \circ s(n^{-\frac{1}{2}})$. This follows from Theorem 6.1.2.◇

We also can apply Theorem 6.1.2 to the symmetrization $\mathcal{T}(V)_{(\gamma,\alpha)}$ which was introduced in Chapter 3. This is the free product $\mathbb{C}\mathbb{L} \star \mathcal{T}(V)$ of algebras divided by the relations $\lambda v = \lambda.v\lambda$ with the coalgebra structure given by

$$\Delta_\gamma(v) = v \otimes \mathrm{d}(v) + \mathbf{1} \otimes v; \ \ \Delta_\gamma(\lambda) = \lambda \otimes \lambda \tag{6.3.1}$$

and

$$\epsilon_\gamma(v) = 0; \ \epsilon_\gamma(\lambda) = 1. \tag{6.3.2}$$

A quantum random variable on $\mathcal{T}(V)_{(\gamma,\alpha)}$ consists of a pair (F, G) where

$$F : V \to \mathcal{A} \text{ and } G : \mathbb{C}\mathbb{L} \to \mathcal{A}$$

are linear and hermitian and such that

$$G(\lambda)F(v) = F(\lambda.v)G(\lambda)$$

for all $\lambda \in \mathbb{L}$, $v \in V$. The coalgebra $\mathcal{T}(V)_\gamma$ is $\mathbb{N}$-graded if we define $\mathrm{l}(M)$ for homogeneous polynomials $M \in \mathcal{T}(V) \subset \mathcal{T}(V)_\gamma$ as before and put $\mathrm{l}(\lambda) = 0$.

6.3.2. THEOREM. *Let $(F_n, G_n)_{n\in\mathbb{N}}$ be an independent sequence of identically distributed quantum random variables on $\mathcal{T}(V)_{(\gamma,\alpha)}$ such that*

$$\varphi(v) = 0 \text{ and } \varphi(\lambda) = 1$$

for all $v \in V$, $\lambda \in \mathbb{L}$. Then the correlations of the processes $(\tilde{S}_{n,t}, R_{n,t})$ with

$$\tilde{S}_{n,t} = \frac{1}{\sqrt{n}} \sum_{l=1}^{[nt]} \tilde{F}_{l,[nt]}; \ \tilde{F}_{l,m}(v) = F_l(v) G_{l+1}(\mathrm{d}_v) \ldots G_m(\mathrm{d}_v) \tag{6.3.3}$$

and

$$R_{n,t}(\lambda) = G_1(\lambda)G_2(\lambda)\ldots G_{[nt]}(\lambda) \tag{6.3.4}$$

converge to the correlations of the symmetric white noise on $\mathcal{T}(V)_{(\gamma,\alpha)}$ with generator g_φ given by

$$g_\varphi(M) = \begin{cases} \varphi(v.\lambda_2 w) & \text{if } M = \lambda_1 v \lambda_2 w \lambda_3 \\ 0 & \text{otherwise} \end{cases}$$

for a monomial M in $\mathcal{T}(V)_{(\gamma,\alpha)}$.

PROOF: An application of Theorem 6.1.3 yields the convergence of $(\tilde{S}_{n,t}, R_{n,t})$ to the white noise with generator g_φ given by $g_\varphi(M) = \varphi(M)$ if $\mathrm{l}(M) = 2$ and $g_\varphi(M) = 0$ if $\mathrm{l}(M) \neq 2$. The condition $\varphi(\lambda) = 1$ gives $\varphi(\lambda_1 b \lambda_2) = \varphi(b)$ for all $\lambda_1, \lambda_2 \in \mathbb{L}$, $b \in \mathcal{T}(V)_{(\gamma,\alpha)}$ by an application of Cauchy-Schwartz inequality. But this means

$$\varphi(\lambda_1 v \lambda_2 w \lambda_3) = \varphi(\lambda_1 \lambda_2 v . \lambda_2 w \lambda_3) = \varphi(v . \lambda_2 w). \diamond$$

Theorems 6.3.1 and 6.3.2 are actually two different aspects of the same thing. For the (only formally defined) sequence $(\tilde{\tilde{F}}_l)_{l\in\mathbb{N}}$ with $\tilde{\tilde{F}}_l = F_l G_{l+1} G_{l+2} \ldots$ is left independent, identically distributed if we regard $\tilde{\tilde{F}}_l$ as quantum random vectors on V. Moreover, g_φ is nothing but $(g_Q)_\alpha$ with $Q(v, w) = \varphi(v^* w)$. For the corresponding invariance principle in the right case this becomes rigorous. In Theorem 6.3.2 $(\tilde{S}_{n,t}, R_{n,t})$ is replaced by $(\check{S}_{n,t}, R_{n,t})$ with

$$\check{S}_{n,t} = \frac{1}{\sqrt{n}} \sum_{l=1}^{[nt]} \check{F}_l; \ \check{F}_l(v) = G_1(\mathrm{d}_v) G_2(\mathrm{d}_v) \ldots G_{l-1}(\mathrm{d}_v) F_l(v).$$

The limiting white noise is the white noise on $\mathcal{T}(V)_{(\alpha,\gamma)}$ with generator $(g_Q)^\alpha$. Moreover, the $\check{F}_l$ form a right independent sequence of quantum random variables on V and by the right version of Theorem 6.3.1 we have that $\check{S}_{n,t}$ converge to the additive right white noise on V with generator g_Q; cf. Section 4.3.

The following construction will illustrate the situation. We treat the case when $V = V_0 \oplus V_0^*$ with $V_0^* = \{v_0^* : v_0 \in V\}$ and $\mathbb{L} = (\mathbb{Z} \times \mathbb{Z})^{\mathrm{twist}}$, $\mathrm{d}(v_0) = (1, 0)$, $(1, 0).v_0 = q v_0$, $(0, 1).v_0 = \bar{q} v_0$, for $v_0 \in V_0$ with $q \in \mathbb{C}_0$; cf. Section 1.6. This is the commutation factor case with $\mathbb{M} = \mathbb{Z}^{\mathrm{inv}}$, $\mathrm{g}(v_0) = 1$ and

$$q((m_1, m_2), n) = q^{m_1 n} \bar{q}^{m_2 n}.$$

In the case when $q = \pm 1$ the invariance condition (4.3.1) on Q is trivial, and if $q \neq \pm 1$ it means

$$Q(v_0, w_0^*) = Q(v_0^*, w_0) = 0$$

for all $v_0, w_0 \in V_0$ and Q is of the form $Q = \begin{pmatrix} Q_1 & 0 \\ 0 & Q_2 \end{pmatrix}$ with positive sesquilinear forms Q_1 and Q_2 on V_0 and V_0^* respectively. We denote the symmetrization of $\mathcal{T}(V)$ by $\mathcal{T}(V)_q$.

We also introduce the 'degenerate' symmetric $*$-bialgebra $\mathcal{T}(V)_0$ which is the free product $\mathbb{C}\mathbb{N} \star \mathcal{T}(V)$ of the $*$-algebras $\mathbb{C}\mathbb{N}$ and $\mathcal{T}(V)$ with the relations $m v_0 = 0$ for all $m \in \mathbb{N}$ and $v_0 \in V_0$. The comultiplication Δ_0 and the counit ε_0 are again given by (6.3.1) and (6.3.2). The quantum random variables on $\mathcal{T}(V)_0$ are pairs (F, G) with

$$G(m) F(v_0) = 0$$

for all $m \in \mathbb{N}$, $v_0 \in V_0$. We have

6.3.3. THEOREM. *Let $(F_n, G_n)_{n\in\mathbb{N}}$ be an independent sequence of identically distributed quantum random variables on $\mathcal{T}(V)_0$ such that*

$$\varphi(v) = 0 \text{ and } \varphi(m) = 1$$

for all $v \in V$, $m \in \mathbb{N}$. Then the correlations of the 'deformed sums' $(\tilde{S}_{n,t}, R_{n,t})$ given by (6.3.3) and (6.3.4) converge to the correlations of the white noise on $\mathcal{T}(V)_0$ with generator g_φ given by

$$g_\varphi(M) = \begin{cases} \varphi(v_0^* w_0) & \text{if } M = m_1 v_0^* w_0 m_2 \\ 0 & \text{otherwise .} \end{cases}$$

PROOF: From Theorem 6.1.2 the sequence $(\tilde{S}_{n,t}, G_{n,t})$ converges to the white noise with generator g_φ where $g_\varphi(M) = \varphi(M)$ if $\mathrm{l}(M) = 2$ and $g_\varphi(M) = 0$ if $\mathrm{l}(M) \neq 2$. The conditions on φ again give $\varphi(m_1 b m_2) = \varphi(b)$ for all $m_1, m_2 \in \mathbb{N}$, $b \in \mathcal{T}(V)$. The relations in $\mathcal{T}(V)_0$ imply that

$$\varphi(m_1 v m_2 w m_3) = \varphi(v m_2 w) = 0$$

unless $m_2 = 0$ and $v \in V_0^*$, $w \in V_0$.$\diamondsuit$

For simplicity, let us now assume that $\dim V_0 = 1$, say V_0 is spanned by x. Via the GNS-construction, the $\mathbb{M}$-even states on $\mathcal{T}(V) = \mathbb{C}\langle x, x^*\rangle$ arise as follows. Let $(D_m)_{m\in\mathbb{Z}}$ be a sequence of pre-Hilbert spaces and let $a \in \mathbf{H}(D)$, $D = \bigoplus_{m\in\mathbb{Z}} D_m$, map D_n to D_{n+1}. For $q \in \mathbb{C}_0$ denote by $b_q = b$ the operator on D which equals multiplication by q^m if restricted to D_m. The situation can be described by the figure

$$\begin{array}{ccccccccccc}
 & & & a & & a & & a & & a & \\
\cdots & \rightarrow & D_{-2} & \rightarrow & D_{-1} & \rightarrow & D_0 & \rightarrow & D_1 & \rightarrow & D_2 & \rightarrow & \cdots \\
 & & \updownarrow q^{-2} & & \updownarrow q^{-1} & & \updownarrow q^{0} & & \updownarrow q^{1} & & \updownarrow q^{2} & & \\
\cdots & \rightarrow & D_{-2} & \rightarrow & D_{-1} & \rightarrow & D_0 & \rightarrow & D_1 & \rightarrow & D_2 & \rightarrow & \cdots \\
 & & & a & & a & & a & & a &
\end{array}$$

For a state Φ_0 given by a unit vector Ω in D_0 the distribution of the quantum random vector a on $V = \mathrm{Lin}\{x, x^*\}$ is $\mathbb{M}$-even. We have the relations

$$ba = qab \text{ and } a^*b = qba^*.$$

Let $\mathcal{A}$ be the infinite tensor product

$$\mathcal{A} = \mathbf{H}(D) \otimes \mathbf{H}(D) \otimes \ldots$$

with respect to the vector state Φ_0 on $\mathbf{H}(D)$. Then $\mathcal{A}$ is the sub-algebra of the complete direct product

$$\mathbf{H}(D)^{\otimes\mathbb{N}} = \mathrm{Lin}\{a_1 \otimes a_2 \otimes \cdots : a_k \in \mathbf{H}(D)\}$$

generated by the elements of the form $a_1 \otimes a_2 \otimes \ldots$ with

$$\Phi_0((a_k - 1)^*(a_k - 1)) = 0$$

but for finitely many k. We put $\Phi = \Phi_0 \otimes \Phi_0 \otimes \dots$ to obtain the quantum probability space $(\mathcal{A}, \Phi)$. Let $\sigma : \mathcal{A} \to \mathcal{A}$ be the shift, i.e.

$$\sigma(a_1 \otimes a_2 \otimes \dots) = \mathbf{1} \otimes a_1 \otimes a_2 \otimes \dots$$

We identify $c \in \mathcal{A}_0$ with $c \otimes \mathbf{1} \otimes \mathbf{1} \otimes \dots \in \mathcal{A}$. We put

$$a^{(k)} = \sigma^{k-1} a \text{ and } b^{(k)} = \sigma^{k-1} b.$$

and for $m \leq n$

$$b_{mn} = b^{(m+1)} b^{(m+2)} \dots b^{(n)}.$$

Next we form three types of sums

$$\Sigma_n = \sum_{l=1}^{n} a^{(l)} b_{l,\infty}, \; \check{\Sigma}_n = \sum_{l=1}^{n} b_{0,l-1} a^{(l)} \text{ and } \tilde{\Sigma}_n = \sum_{l=1}^{n} a^{(l)} b_{ln}.$$

We also form four types of increments for $m \leq n$

$$\Sigma_{mn} = \sum_{l=m+1}^{n} a^{(l)} b_{l,\infty} = \Sigma_n - \Sigma_m,$$

$$\check{\Sigma}_{mn} = \sum_{l=m+1}^{n} b_{0,l-1} a^{(l)} = \check{\Sigma}_n - \check{\Sigma}_m,$$

$$\tilde{\Sigma}_{mn} = \sum_{l=m+1}^{n} a^{(l)} b_{ln} = \tilde{\Sigma}_n - \tilde{\Sigma}_m b_{mn}$$

and

$$\tilde{\check{\Sigma}}_{mn} = \sum_{l=m+1}^{n} b_{m,l-1} a^{(l)} = (b_{0,m-1})^{-1} (\check{\Sigma}_n - \check{\Sigma}_m).$$

The family $(\Sigma_{mn})_{0 \leq m \leq n}$ with the obvious order is left q-independent whereas $(\check{\Sigma}_{mn})$ is right q-independent. Moreover, $(\tilde{\Sigma}_{mn})$ and $(\tilde{\check{\Sigma}}_{mn})$ are the symmetrizations of (Σ_{mn}) and $(\check{\Sigma}_{mn})$ respectively, i.e.

$$\Sigma_{mn} = \tilde{\Sigma}_{mn} b_{n,\infty} \text{ and } \check{\Sigma}_{mn} = b_{0,m-1} \tilde{\check{\Sigma}}_{mn}.$$

Σ_{mn} and $\check{\Sigma}_{mn}$ are discrete versions of an additive left q-white noise and an additive right q-white noise respectively. The symmetrizations are discrete versions of white noise on $\mathcal{T}(V)_{(\gamma,\alpha)}$ and $\mathcal{T}(V)_{(\alpha,\gamma)}$. We have

$$\tilde{\Sigma}_m \tilde{\Sigma}_n = \tilde{\Sigma}_m (\tilde{\Sigma}_m b_{mn} + \tilde{\Sigma}_{mn}) = \tilde{\Sigma}_n \tilde{\Sigma}_m$$

so that $X_m = \tilde{\Sigma}_m + (\tilde{\Sigma}_m)^*$ forms a commuting family of symmetric operators, and if we assume that $a + a^*$ is a self-adjoint operator we can conclude via the spectral theory that X_m, in the state Φ, forms a classical sequence of random variables. These random variables will not be ordinary sums of independent identically distributed random variables but the deformed increments $X_{mn} = X_n - X_m b_{mn}$ (where we assumed $q \in \mathbb{R}$ for simplicity!) are independent in the sense of non-commutative probability theory. In the state Φ, they give rise to a sequence of independent classical random variables. However, the families (X_{mn}) and (X_m) do not commute in general, and it will not always be possible to realize them simultaneously, in the state Φ, as classical random variables.

Our discrete processes also satisfy certain quantum stochastic difference equations. For instance,

$$\tilde{\Sigma}_{n+1} - \tilde{\Sigma}_n = \tilde{\Sigma}_n b_{n,n+1} + \tilde{\Sigma}_{n,n+1} - \tilde{\Sigma}_n = \tilde{\Sigma}_n (b^{(n+1)} - \mathrm{id}) + a^{(n+1)}.$$

The basic example is when $D_0 = \mathbb{C}$, $D_1 = \mathbb{C}$, $D_n = \{0\}$ for $n \notin \{0,1\}$, $a\lceil D_0 = \mathrm{id}$ and $\Omega = 1$. Then $D = \mathbb{C}^2$, $a = \begin{pmatrix} 0 & 0 \\ 1 & 0 \end{pmatrix}$, $b = \begin{pmatrix} 1 & 0 \\ 0 & q \end{pmatrix}$. In this case, the above difference equation becomes

$$\tilde{\Sigma}_{n+1} - \tilde{\Sigma}_n = (q-1)\tilde{\Sigma}_n (aa^*)^{(n+1)} + a^{(n+1)}$$

which can be regarded as the discrete version of the quantum stochastic differential equation for quantum Azéma noise. The classical sequence (X_n) is the classical Markov chain that led P.A. Meyer and K.R. Parthasarathy to the quantum stochastic differential equation of the Azéma martingale; cf. [58, 61].

Theorem 6.3.2 tells us that the normalized sums $\frac{1}{\sqrt{n}}\tilde{\Sigma}_{[nt]}$ and $\frac{1}{\sqrt{n}}\check{\Sigma}_{[nt]}$ converge to 2-dimensional Azéma noise (L_t, L_t^*) and to the interpolation $(\check{L}_t, \check{L}_t^*)$ of 2-dimensional quantum Brownian motion respectively with the covariance matrix given by

$$Q = \begin{pmatrix} \|a^*\Omega\|^2 & 0 \\ 0 & \|a\Omega\|^2 \end{pmatrix}.$$

Moreover, $\frac{1}{\sqrt{n}}\Sigma_{[nt]}$ converges to $(L_t\Gamma_{t,\infty}(b_q), L_t^*\Gamma_{t,\infty}(b_{\bar{q}}))$ by Theorem 6.3.1.

We conclude with the treatment of the case $q = 0$. Assume that we are given a quantum probability space $(\mathcal{A}_0, \Phi_0)$ and elements $a, b \in \mathcal{A}_0$ such that $ba = 0$, $b^* = b$, $\Phi_0(b) = \Phi_0(b^2) = 1$. Then we must have $\Phi_0(a^2) = \Phi_0((a^*)^2) = \Phi_0(aa^*) = 0$. As before the sums Σ_n, $\tilde{\Sigma}_n$ and $\check{\Sigma}_n$ can be formed, and an application of Theorem 6.3.3 yields that $\frac{1}{\sqrt{n}}\tilde{\Sigma}_{[nt]}$ converges to the solution L_t of

$$F_t = -\int_0^t F_r \mathrm{d}\Lambda_r + \Phi_0(a^*a)^{\frac{1}{2}} A_t,$$

$\frac{1}{\sqrt{n}}\Sigma_{[nt]}$ converges to $L_t\Gamma_{t,\infty}(0)$ and $\frac{1}{\sqrt{n}}\check{\Sigma}_{[nt]}$ converges to $\check{L}_t$ given by

$$\check{L}_t = \Phi_0(a^*a)^{\frac{1}{2}} \int_0^t \Gamma_r(0) \mathrm{d}A_r.$$

REFERENCES

[1] Abe, E.: Hopf algebras. Cambridge: Cambridge University Press 1980

[2] Accardi, L., Lu, Y.G.: Quantum central limit theorems for weakly dependent maps. Preprint, Rome 1990

[3] Accardi, L., Parthasarathy, K.R.: A martingale characterization of canonical commutation and anti-commutation relations. J. Funct. Anal. **77**, 211-231 (1988)

[4] Accardi, L., Frigerio, A., Lewis, J.T.: Quantum stochastic processes. Publ. RIMS Kyoto Univ. **18**, 97-133 (1982)

[5] Accardi, L., Schürmann, M., Waldenfels, W.v.: Quantum independent increment processes on superalgebras. Math. Z. **198**, 451-477 (1988)

[6] Applebaum, D.: The strong Markov property for Fermion Brownian motion. J. Funct. Anal. **65**, 273-291 (1986)

[7] Applebaum, D., Hudson, R.L.: Fermion Ito's formula and stochastic evolutions. Commun. Math. Phys. **96**, 473-496 (1984)

[8] Araki, H.: Factorizable representations of current algebra. Publ. RIMS Kyoto Univ. **5**, 361-422 (1970)

[9] Araki, H., Woods, E.J.: Complete Boolean algebras of type I factors. Publ. RIMS Kyoto Univ. **2**, 157-242 (1967)

[10] Azéma, J.: Sur les fermes aleatoire. In: Azéma, J., Yor, M. (eds.) Sem. Probab. XIX, Strasbourg 1983/84. (Lect. Notes Math., vol. 1223). Berlin Heidelberg New York: Springer 1985

[11] Barchielli, A., Lupieri, G.: Convolution semigroups in quantum probability and quantum stochastic calculus. In: Accardi, L., Waldenfels, W.v. (eds.) Quantum probability and applications IV. Proceedings, Rome 1987. (Lect. Notes Math., vol. 1396).Berlin Heidelberg New York: Springer 1989

[12] Barchielli, A., Lupieri, G.: Semigroups of positive-definite maps on $*$-bialgebras. Preprint, Milano 1991

[13] Barchielli, A., Holevo, A.S., Lupieri, G.: An analogue of Hunt's representation theorem in quantum probability. Preprint, Milano 1990

[14] Barnett, C., Streater, R.F., Wilde, I.F.: The Ito-Clifford integral. J. Funct. Anal. **48**, 172-212 (1982)

[15] Barnett, C., Streater, R.F., Wilde, I.F.: Quasi-free quantum stochastic integrals for the CAR and CCR. J. Funct. Anal. **52**, 19-47 (1983)

[16] Belavkin, V.P.: A quantum nonadapted Ito formula and stochastic analysis in Fock space. J. Funct. Anal. **102**, 414-447 (1991)

[17] Berg, C., Christensen, J.P.R., Ressel, P.: Harmonic analysis on semi-groups. Berlin Heidelberg New York: Springer 1984

[18] Bourbaki, N.: Elements of mathematics. Algebra. Chap I-III. Paris: Hermann 1973

[19] Bozejko, M., Speicher, R.: Examples of generalized white noises. Preprint, Heidelberg 1990

[20] Bozejko, M., Speicher, R.: An example of a generalized Brownian motion. Commun. Math. Phys. **137**, 519-531 (1991)

[21] Chung, K.L.: A course in probability theory. New York: Harcourt, Brace and World 1968

[22] Cockcroft, A.M., Hudson, R.L.: Quantum mechanical Wiener processes. J. Multivariate Anal. **7**, 107-124 (1977)

[23] Cushen, C.D., Hudson, R.L.: A quantum-mechanical central limit theorem. J. Appl. Prob. **8**, 454-469 (1971)

[24] Dixmier, J.: Les algebres d'opérateur dans l'espace Hilbertienne. Paris: Gauthier-Villar 1969

[25] Drinfel'd, V.G.: Quantum groups. Proceedings of the International Congress of Mathematicians, Berkeley 1986

[26] Emery, M.: On the Azéma martingales. In: Sem. Probab. XXIII, Strasbourg. (Lect. Notes Math., vol. 1372). Berlin Heidelberg New York: Springer 1989

[27] Frigerio, A., Maassen, H.: Quantum Poisson processes and dilations of dynamical semigroups. Probab. Th. Rel. Fields **83**, 489-508 (1989)

[28] Giri, N., Waldenfels, W.v.: An algebraic central limit theorem. Z. Wahrsch. verw. Gebiete **42**, 129-134 (1978)

[29] Glockner, P.: Quantum stochastic differential equations on $*$-bialgebras. Math. Proc. Cambridge Phil. Soc. **109**, 571-595 (1991)

[30] Glockner, P., Waldenfels, W.v.: The relations of the non-commutative coefficient algebra of the unitary group. SFB-Preprint Nr. 460, Heidelberg 1988

[31] Glockner, P., Schürmann, M., Speicher, R.: Realization of free white noise. Arch. Math. **58**, 407-416 (1992)

[32] Guichardet, A.: Symmetric Hilbert spaces and related topics. (Lect. Notes Math., vol. 261). Berlin Heidelberg New York: Springer 1972

[33] Hegerfeldt, G.C.: Noncommutative analogs of probabilistic notions and results. J. Funct. Anal. **64**, 436-456 (1985)

[34] Hewitt, E., Ross, K.A.: Abstract harmonic analysis, vols. I,II. Berlin Heidelberg New York: Springer 1970

[35] Heyer, H.: Probability measures on locally compact groups. Berlin Heidelberg New York: Springer 1977

[36] Hida, T.: Brownian motion. Berlin Heidelberg New York: Springer 1980

[37] Holevo, A.S.: An analog of the Ito decomposition for multiplicative processes with values in a Lie group. In: Accardi, L., Waldenfels, W.v. (eds.) Quantum probability and applications V. Proceedings, Heidelberg 1988. (Lect. Notes Math., vol. 1442). Berlin Heidelberg New York: Springer 1990

[38] Hudson, R.L.: A quantum-mechanical central limit theorem in the anti-commuting case. J. Appl. Prob. **10**, 502-509 (1973)

[39] Hudson, R.L., Lindsay, J.M.: A non-commutative martingale representation theorem for non-Fock quantum Brownian motion. J. Funct. Anal. **61**, 202-221 (1984)

[40] Hudson, R.L., Lindsay, J.M.: On characterizing quantum stochastic evolutions. Math. Proc. Philos. Soc. **102**, 363-369 (1987)

[41] Hudson, R.L., Parthasarathy, K.R.: Quantum Ito's formula and stochastic evolutions. Commun. Math. Phys. **93**, 301-323 (1984)

[42] Hudson, R.L., Parthasarathy, K.R.: Stochastic dilations of uniformly continuous completely positive semigroups. Acta Appl. Mathematicae **2**, 353-398 (1984)

[43] Hudson, R.L., Parthasarathy, K.R.: Unification of fermion and boson stochastic calculus. Commun. Math. Phys. **104**, 457-470 (1986)

[44] Hunt, G.A.: Semi-groups of measures on Lie-groups. Trans. Amer. Math. Soc. **81**, 264-293 (1956)

[45] Jacobson, N.: Lie algebras. New York, London: Wiley 1962

[46] Kac, V.G.: Lie superalgebras. Adv. in Math. **26**, 8-96 (1977)

[47] Kümmerer, B.: Markov dilations and non-commutative Poisson processes. Preprint, Tübingen 1989

[48] Kümmerer, B., Speicher, R.: Stochastic integration on the Cuntz algebra O_∞. J. Funct. Anal. **103**, 372-408 (1992)

[49] Lindsay, J.M.: On set convolutions and integral-sum kernel operators. Preprint, Nottingham 1989

[50] Lindsay, J.M., Parthasarathy, K.R.: Cohomology of power sets with applications in quantum probability. Commun. Math. Phys. **124**, 337-364 (1989)

[51] Maassen, H.: Quantum Markov processes on Fock space described by integral kernels. In: Accardi, L., Waldenfels, W.v. (eds.) Quantum probability and applications II. Proceedings, Heidelberg 1984. (Lect. Notes Math., vol. 1136). Berlin Heidelberg New York: Springer 1985

[52] Maassen, H.: Theoretical concepts in quantum probability: Quantum Markov processes. Preprint, Nijmegen 1988

[53] Majid, S.: Quasitriangular Hopf algebras and Yang-Baxter equations. Int. J. Modern Physics **A 5(1)**, 1-91 (1990)

[54] Majid, S.: Braided groups and braid statistics. Preprint, Cambridge 1991

[55] Majid, S.: Cross products by braided groups and bosonization. Preprint, Cambridge 1991

[56] Mathon, D., Streater, R.F.: Infinitely divisible representations of Clifford algebras. Z. Wahrsch. verw. Gebiete **20**, 308-316 (1971)

[57] Meyer, P.A.: Fock spaces in classical and non-commutative probability, Chap. I-IV. Publication de l'IRMA, Strasbourg 1989

[58] Meyer, P.A.: New examples of chaotic representation. Preprint, Strasbourg 1988

[59] Parthasarathy, K.R.: Some remarks on the integration of Schrödinger equation using the quantum stochastic calculus. In: Accrdi, L., Waldenfels, v.W. (eds.) Quantum probability and applications II. Proceedings, Heidelberg 1984. (Lect. Notes Math., vol. 1136). Berlin Heidelberg New York: Springer 1985

[60] Parthasarathy, K.R.: A unified approach to classical, bosonic and fermionic Brownian motions. Colloques Paul Lévy sur les processus stochastiques. Astérisque **157/158**, 303-320 (1988)

[61] Parthasarathy, K.R.: Azéma martingales and quantum stochastic calculus. In: Bahadur, R.R. (ed.) Proc. R.C. Bose Memorial Symposium. New Delhi: Wiley Eastern 1990

[62] Parthasarathy, K.R.: An introduction to quantum stochastic calculus. Basel Boston Berlin: Birkhäuser 1992

[63] Parthasarathy, K.R., Schmidt, K.: Positive definite kernels, continuous tensor products, and central limit theorems of probability theory. (Lect. Notes Math., vol. 272). Berlin Heidelberg New York: Springer 1972

[64] Parthasarathy, K.R., Sinha, K.B.: Stochastic integral representation of bounded quantum martingales in Fock space. J. Funct. Anal. **67**, 126-151 (1986)

[65] Parthasarathy, K.R., Sinha, K.B.: Representation of a class of quantum martingales II. In: Accardi, L., Waldenfels, v.W. (eds.) Quantum probability and applications III. Proceedings, Oberwolfach 1978. (Lect. Notes Math., vol. 1303). Berlin Heidelberg New York: Springer 1988

[66] Parthasarathy, K.R., Sinha, K.B.: Unification of quantum noise in Fock spaces. In: Accardi, L. (ed.) Quantum probability VI. Singapore New Jersey London Hong Kong: World Scientific 1991

[67] Powers, R.: Self-adjoint algebras of unbounded operators. Commun. Math. Phys. **21**, 85-124 (1971)

[68] Prin, J.: Verallgemeinertes Weißes Rauschen und nichtkommutative stochastische Integration. Diplomarbeit, Tübingen 1989

[69] Schoenberg, I.J.: Metric spaces and positive definite functions. Trans. Amer. Math. Soc. **44**, 522-536 (1938)

[70] Schürmann, M.: Positive and conditionally positive linear functionals on coalgebras. In: Accardi, L., Waldenfels, v.W. (eds.) Quantum probability and applications II. Proceedings, Heidelberg 1984. (Lect. Notes Math., vol. 1136). Berlin Heidelberg New York: Springer 1985

[71] Schürmann, M.: A class of representations of involutive bialgebras. Math. Proc. Camb. Phil. Soc. **107**, 149-175 (1990)

[72] Schürmann, M.: Noncommutative stochastic processes with independent and stationary increments satisfy quantum stochastic differential equations. Probab. Th. Rel. Fields **84**, 473-490 (1990)

[73] Schürmann, M.: Gaussian states on bialgebras. In: Accardi, L., Waldenfels, v.W. (eds.) Quantum probability and applications V. Proceedings, Heidelberg 1988. (Lect. Notes Math., vol. 1442). Berlin Heidelberg New York: Springer 1990

[74] Schürmann, M.: Quantum stochastic processes with independent additive increments. J. Multivariate Anal. **38**, 15-35 (1991)

[75] Schürmann, M.: White noise on involutive bialgebras. In: Accardi, L. (ed.) Quantum probability VI. Singapore, New Jersey, London, Hong Kong: World Scientific 1991

[76] Schürmann, M.: The Azéma martingales as components of quantum independent increment processes. In: Azéma, J., Meyer, P.A., Yor, M. (eds) Séminaire de probabilités XXV. (Lect. Notes Math., vol. 1485). Berlin Heidelberg New York: Springer 1991

[77] Schürmann, M.: Quantum q-white noise and a q-central limit theorem. Commun. Math. phys. **140**, 589-615 (1991)

[78] Schürmann, M.: Realization of unitary q-white noise on Fock space. In: Heyer, H. (ed.) Probability measures on groups X. Proceedings, Oberwolfach 1990. New York: Plenum Press 1991

[79] Schürmann, M.: The Azéma martingales as central limits. To appear in: Quantum probability VII. Proceedings, New Delhi 1991

[80] Schürmann, M., Waldenfels, W.v.: A central limit theorem on the free Lie group. In: Accardi, L., Waldenfels, v.W. (eds.) Quantum probability and applications

III. proceedings, Oberwolfach 1987. (Lect. Notes Math., vol. 1303). Berlin Heidelberg New York: Springer 1988

[81] Skorokhod, A.V.: Operator stochastic differential equations and stochastic semigroups. Russian Math. Surveys **37**, 177-204 (1982)

[82] Speicher, R.: A new example of 'independence' and 'white noise'. Probab. Th. Rel. Fields **84**, 141-159 (1990)

[83] Speicher, R.: Stochastic integration on the full Fock space with the help of a kernel calculus. Publ. RIMS Kyoto Univ. **27**, 149-184 (1991)

[84] Speicher, R.: A non-commutative central limit theorem. Math. Z. **209**, 55-66 (1992)

[85] Streater, R.F.: Current commutation relations, continuous tensor products and infinitely divisible group representations. In: Jost, R. (ed.) Local quantum theory. London New York: Academic Press 1969

[86] Streater, R.F.: Infinitely divisible representations of Lie algebras. Z. Wahrsch. verw. Gebiete **19**, 67-80 (1971)

[87] Sweedler, M.E.: Hopf algebras. New York: Benjamin 1969

[88] Voiculescu, D.: Symmetries of some reduced free product C*-algebras. In: Araki, H., Moore, C.C., Stratila, S., Voiculescu, D. (eds.) Operator algebras and their connection with topology and ergodic theory. Proceedings, Busteni 1983. (Lect. Notes Math., vol. 1132). Berlin Heidelberg New York: Springer 1985

[89] Voiculescu, D.: Dual algebraic structures on operator algebras related to free products. J. Operator Theory **17**, 85-98 (1987)

[90] Waldenfels, W.v.: An algebraic central limit theorem in the anti-commuting case. Z. Wahrsch. verw. Gebiete **42**, 135-140 (1978)

[91] Waldenfels, W.v.: Positive and conditionally positive sesquilinear forms on anti-cocommutative coalgebras. In: Heyer, H. (ed.) Probability measures on groups VII. Proceedings, Oberwolfach 1983 (Lect. Notes Math., vol. 1064). Berlin Heidelberg New York: Springer 1983

[92] Waldenfels, W.v.: Ito solution of the linear quantum stochastic differential equation describing light emission and absorption. In: Accardi, L., Frigerio, A., Gorini, V. (eds.) Quantum probability and applications to the theory of irreversible processes. Proceedings, Villa Mondragone 1982. (Lect. Notes Math., vol. 1055). Berlin Heidelberg New York: Springer 1984

[93] Woronowicz, S.L.: Twisted $SU(2)$ group. An example of a non-commutative differential calculus. Publ. RIMS, Kyoto Univ. **23**, 117-181 (1987)

[94] Woronowicz, S.L.: Compact matrix pseudo groups. Commun. Math. Phys. **111**, 613-665 (1987)

SUBJECT INDEX

R

S

T

U

V

W

Printing: Druckhaus Beltz, Hemsbach
Binding: Buchbinderei Schäffer, Grünstadt

Springer-Verlag and the Environment

We at Springer-Verlag firmly believe that an international science publisher has a special obligation to the environment, and our corporate policies consistently reflect this conviction.

We also expect our business partners – paper mills, printers, packaging manufacturers, etc. – to commit themselves to using environmentally friendly materials and production processes.

The paper in this book is made from low- or no-chlorine pulp and is acid free, in conformance with international standards for paper permanency.

Lecture Notes in Mathematics

For information about Vols. 1–1340
please contact your bookseller or Springer-Verlag

Vol. 1387: M. Petković, Iterative Methods for Simultaneous Inclusion of Polynomial Zeros. X, 263 pages. 1989.

Vol. 1388: J. Shinoda, T.A. Slaman, T. Tugué (Eds.), Mathematical Logic and Applications. Proceedings, 1987. V, 223 pages. 1989.

Vol. 1000: Second Edition. H. Hopf, Differential Geometry in the Large. VII, 184 pages. 1989.

Vol. 1389: E. Ballico, C. Ciliberto (Eds.), Algebraic Curves and Projective Geometry. Proceedings, 1988. V, 288 pages. 1989.

Vol. 1390: G. Da Prato, L. Tubaro (Eds.), Stochastic Partial Differential Equations and Applications II. Proceedings, 1988. VI, 258 pages. 1989.

Vol. 1391: S. Cambanis, A. Weron (Eds.), Probability Theory on Vector Spaces IV. Proceedings, 1987. VIII, 424 pages. 1989.

Vol. 1392: R. Silhol, Real Algebraic Surfaces. X, 215 pages. 1989.

Vol. 1393: N. Bouleau, D. Feyel, F. Hirsch, G. Mokobodzki (Eds.), Séminaire de Théorie du Potentiel Paris, No. 9. Proceedings. VI, 265 pages. 1989.

Vol. 1394: T.L. Gill, W.W. Zachary (Eds.), Nonlinear Semigroups, Partial Differential Equations and Attractors. Proceedings, 1987. IX, 233 pages. 1989.

Vol. 1395: K. Alladi (Ed.), Number Theory, Madras 1987. Proceedings. VII, 234 pages. 1989.

Vol. 1396: L. Accardi, W. von Waldenfels (Eds.), Quantum Probability and Applications IV. Proceedings, 1987. VI, 355 pages. 1989.

Vol. 1397: P.R. Turner (Ed.), Numerical Analysis and Parallel Processing. Seminar, 1987. VI, 264 pages. 1989.

Vol. 1398: A.C. Kim, B.H. Neumann (Eds.), Groups – Korea 1988. Proceedings. V, 189 pages. 1989.

Vol. 1399: W.-P. Barth, H. Lange (Eds.), Arithmetic of Complex Manifolds. Proceedings, 1988. V, 171 pages. 1989.

Vol. 1400: U. Jannsen. Mixed Motives and Algebraic K-Theory. XIII, 246 pages. 1990.

Vol. 1401: J. Steprans, S. Watson (Eds.), Set Theory and its Applications. Proceedings, 1987. V, 227 pages. 1989.

Vol. 1402: C. Carasso, P. Charrier, B. Hanouzet, J.-L. Joly (Eds.), Nonlinear Hyperbolic Problems. Proceedings, 1988. V, 249 pages. 1989.

Vol. 1403: B. Simeone (Ed.), Combinatorial Optimization. Seminar, 1986. V, 314 pages. 1989.

Vol. 1404: M.-P. Malliavin (Ed.), Séminaire d´AIgèbre Paul Dubreil et Marie-Paul Malliavin. Proceedings, 1987–1988. IV, 410 pages. 1989.

Vol. 1405: S. Dolecki (Ed.), Optimization. Proceedings, 1988. V, 223 pages. 1989. Vol. 1406: L. Jacobsen (Ed.), Analytic Theory of Continued Fractions III. Proceedings, 1988. VI, 142 pages. 1989.

Vol. 1407: W. Pohlers, Proof Theory. VI, 213 pages. 1989.

Vol. 1408: W. Lück, Transformation Groups and Algebraic K-Theory. XII, 443 pages. 1989.

Vol. 1409: E. Hairer, Ch. Lubich, M. Roche. The Numerical Solution of Differential-Algebraic Systems by Runge-Kutta Methods. VII, 139 pages. 1989.

Vol. 1410: F.J. Carreras, O. Gil-Medrano, A.M. Naveira (Eds.), Differential Geometry. Proceedings, 1988. V, 308 pages. 1989.

Vol. 1411: B. Jiang (Ed.), Topological Fixed Point Theory and Applications. Proceedings. 1988. VI, 203 pages. 1989.

Vol. 1412: V.V. Kalashnikov, V.M. Zolotarev (Eds.), Stability Problems for Stochastic Models. Proceedings, 1987. X, 380 pages. 1989.

Vol. 1413: S. Wright, Uniqueness of the Injective III_1 Factor. III, 108 pages. 1989.

Vol. 1414: E. Ramirez de Arellano (Ed.), Algebraic Geometry and Complex Analysis. Proceedings, 1987. VI, 180 pages. 1989.

Vol. 1415: M. Langevin, M. Waldschmidt (Eds.), Cinquante Ans de Polynômes. Fifty Years of Polynomials. Proceedings, 1988. IX, 235 pages.1990.

Vol. 1416: C. Albert (Ed.), Géométrie Symplectique et Mécanique. Proceedings, 1988. V, 289 pages. 1990.

Vol. 1417: A.J. Sommese, A. Biancofiore, E.L. Livorni (Eds.), Algebraic Geometry. Proceedings, 1988. V, 320 pages. 1990.

Vol. 1418: M. Mimura (Ed.), Homotopy Theory and Related Topics. Proceedings, 1988. V, 241 pages. 1990.

Vol. 1419: P.S. Bullen, P.Y. Lee, J.L. Mawhin, P. Muldowney, W.F. Pfeffer (Eds.), New Integrals. Proceedings, 1988. V, 202 pages. 1990.

Vol. 1420: M. Galbiati, A. Tognoli (Eds.), Real Analytic Geometry. Proceedings, 1988. IV, 366 pages. 1990.

Vol. 1421: H.A. Biagioni, A Nonlinear Theory of Generalized Functions, XII, 214 pages. 1990.

Vol. 1422: V. Villani (Ed.), Complex Geometry and Analysis. Proceedings, 1988. V, 109 pages. 1990.

Vol. 1423: S.O. Kochman, Stable Homotopy Groups of Spheres: A Computer-Assisted Approach. VIII, 330 pages. 1990.

Vol. 1424: F.E. Burstall, J.H. Rawnsley, Twistor Theory for Riemannian Symmetric Spaces. III, 112 pages. 1990.

Vol. 1425: R.A. Piccinini (Ed.), Groups of Self-Equivalences and Related Topics. Proceedings, 1988. V, 214 pages. 1990.

Vol. 1426: J. Azéma, P.A. Meyer, M. Yor (Eds.), Séminaire de Probabilités XXIV, 1988/89. V, 490 pages. 1990.

Vol. 1427: A. Ancona, D. Geman, N. Ikeda, École d'Eté de Probabilités de Saint Flour XVIII, 1988. Ed.: P.L. Hennequin. VII, 330 pages. 1990.

Vol. 1428: K. Erdmann, Blocks of Tame Representation Type and Related Algebras. XV. 312 pages. 1990.

Vol. 1429: S. Homer, A. Nerode, R.A. Platek, G.E. Sacks, A. Scedrov, Logic and Computer Science. Seminar, 1988. Editor: P. Odifreddi. V, 162 pages. 1990.

Vol. 1430: W. Bruns, A. Simis (Eds.), Commutative Algebra. Proceedings. 1988. V, 160 pages. 1990.

Vol. 1431: J.G. Heywood, K. Masuda, R. Rautmann, V.A. Solonnikov (Eds.), The Navier-Stokes Equations – Theory and Numerical Methods. Proceedings, 1988. VII, 238 pages. 1990.

Vol. 1432: K. Ambos-Spies, G.H. Müller, G.E. Sacks (Eds.), Recursion Theory Week. Proceedings, 1989. VI, 393 pages. 1990.

Vol. 1433: S. Lang, W. Cherry, Topics in Nevanlinna Theory. II, 174 pages.1990.

Vol. 1434: K. Nagasaka, E. Fouvry (Eds.), Analytic Number Theory. Proceedings, 1988. VI, 218 pages. 1990.

Vol. 1435: St. Ruscheweyh, E.B. Saff, L.C. Salinas, R.S. Varga (Eds.), Computational Methods and Function Theory. Proceedings, 1989. VI, 211 pages. 1990.

Vol. 1436: S. Xambó-Descamps (Ed.), Enumerative Geometry. Proceedings, 1987. V, 303 pages. 1990.

Vol. 1437: H. Inassaridze (Ed.), K-theory and Homological Algebra. Seminar, 1987–88. V, 313 pages. 1990.

Vol. 1438: P.G. Lemarié (Ed.) Les Ondelettes en 1989. Seminar. IV, 212 pages. 1990.

Vol. 1439: E. Bujalance, J.J. Etayo, J.M. Gamboa, G. Gromadzki. Automorphism Groups of Compact Bordered Klein Surfaces: A Combinatorial Approach. XIII, 201 pages. 1990.

Vol. 1440: P. Latiolais (Ed.), Topology and Combinatorial Groups Theory. Seminar, 1985–1988. VI, 207 pages. 1990.

Vol. 1441: M. Coornaert, T. Delzant, A. Papadopoulos. Géométrie et théorie des groupes. X, 165 pages. 1990.

Vol. 1442: L. Accardi, M. von Waldenfels (Eds.), Quantum Probability and Applications V. Proceedings, 1988. VI, 413 pages. 1990.

Vol. 1443: K.H. Dovermann, R. Schultz, Equivariant Surgery Theories and Their Periodicity Properties. VI, 227 pages. 1990.

Vol. 1444: H. Korezlioglu, A.S. Ustunel (Eds.), Stochastic Analysis and Related Topics VI. Proceedings, 1988. V, 268 pages. 1990.

Vol. 1445: F. Schulz, Regularity Theory for Quasilinear Elliptic Systems and – Monge Ampère Equations in Two Dimensions. XV, 123 pages. 1990.

Vol. 1446: Methods of Nonconvex Analysis. Seminar, 1989. Editor: A. Cellina. V, 206 pages. 1990.

Vol. 1447: J.-G. Labesse, J. Schwermer (Eds), Cohomology of Arithmetic Groups and Automorphic Forms. Proceedings, 1989. V, 358 pages. 1990.

Vol. 1448: S.K. Jain, S.R. López-Permouth (Eds.), Non-Commutative Ring Theory. Proceedings, 1989. V, 166 pages. 1990.

Vol. 1449: W. Odyniec, G. Lewicki, Minimal Projections in Banach Spaces. VIII, 168 pages. 1990.

Vol. 1450: H. Fujita, T. Ikebe, S.T. Kuroda (Eds.), Functional-Analytic Methods for Partial Differential Equations. Proceedings, 1989. VII, 252 pages. 1990.

Vol. 1451: L. Alvarez-Gaumé, E. Arbarello, C. De Concini, N.J. Hitchin, Global Geometry and Mathematical Physics. Montecatini Terme 1988. Seminar. Editors: M. Francaviglia, F. Gherardelli. IX, 197 pages. 1990.

Vol. 1452: E. Hlawka, R.F. Tichy (Eds.), Number-Theoretic Analysis. Seminar, 1988–89. V, 220 pages. 1990.

Vol. 1453: Yu.G. Borisovich, Yu.E. Gliklikh (Eds.), Global Analysis – Studies and Applications IV. V, 320 pages. 1990.

Vol. 1454: F. Baldassari, S. Bosch, B. Dwork (Eds.), p-adic Analysis. Proceedings, 1989. V, 382 pages. 1990.

Vol. 1455: J.-P. Françoise, R. Roussarie (Eds.), Bifurcations of Planar Vector Fields. Proceedings, 1989. VI, 396 pages. 1990.

Vol. 1456: L.G. Kovács (Ed.), Groups – Canberra 1989. Proceedings. XII, 198 pages. 1990.

Vol. 1457: O. Axelsson, L.Yu. Kolotilina (Eds.), Preconditioned Conjugate Gradient Methods. Proceedings, 1989. V, 196 pages. 1990.

Vol. 1458: R. Schaaf, Global Solution Branches of Two Point Boundary Value Problems. XIX, 141 pages. 1990.

Vol. 1459: D. Tiba, Optimal Control of Nonsmooth Distributed Parameter Systems. VII, 159 pages. 1990.

Vol. 1460: G. Toscani, V. Boffi, S. Rionero (Eds.), Mathematical Aspects of Fluid Plasma Dynamics. Proceedings, 1988. V, 221 pages. 1991.

Vol. 1461: R. Gorenflo, S. Vessella, Abel Integral Equations. VII, 215 pages. 1991.

Vol. 1462: D. Mond, J. Montaldi (Eds.), Singularity Theory and its Applications. Warwick 1989, Part I. VIII, 405 pages. 1991.

Vol. 1463: R. Roberts, I. Stewart (Eds.), Singularity Theory and its Applications. Warwick 1989, Part II. VIII, 322 pages. 1991.

Vol. 1464: D. L. Burkholder, E. Pardoux, A. Sznitman, Ecole d'Eté de Probabilités de Saint- Flour XIX-1989. Editor: P. L. Hennequin. VI, 256 pages. 1991.

Vol. 1465: G. David, Wavelets and Singular Integrals on Curves and Surfaces. X, 107 pages. 1991.

Vol. 1466: W. Banaszczyk, Additive Subgroups of Topological Vector Spaces. VII, 178 pages. 1991.

Vol. 1467: W. M. Schmidt, Diophantine Approximations and Diophantine Equations. VIII, 217 pages. 1991.

Vol. 1468: J. Noguchi, T. Ohsawa (Eds.), Prospects in Complex Geometry. Proceedings, 1989. VII, 421 pages. 1991.

Vol. 1469: J. Lindenstrauss, V. D. Milman (Eds.), Geometric Aspects of Functional Analysis. Seminar 1989-90. XI, 191 pages. 1991.

Vol. 1470: E. Odell, H. Rosenthal (Eds.), Functional Analysis. Proceedings, 1987-89. VII, 199 pages. 1991.

Vol. 1471: A. A. Panchishkin, Non-Archimedean L-Functions of Siegel and Hilbert Modular Forms. VII, 157 pages. 1991.

Vol. 1472: T. T. Nielsen, Bose Algebras: The Complex and Real Wave Representations. V, 132 pages. 1991.

Vol. 1473: Y. Hino, S. Murakami, T. Naito, Functional Differential Equations with Infinite Delay. X, 317 pages. 1991.

Vol. 1474: S. Jackowski, B. Oliver, K. Pawałowski (Eds.), Algebraic Topology, Poznań 1989. Proceedings. VIII, 397 pages. 1991.

Vol. 1475: S. Busenberg, M. Martelli (Eds.), Delay Differential Equations and Dynamical Systems. Proceedings, 1990. VIII, 249 pages. 1991.

Vol. 1476: M. Bekkali, Topics in Set Theory. VII, 120 pages. 1991.

Vol. 1477: R. Jajte, Strong Limit Theorems in Noncommutative L_2-Spaces. X, 113 pages. 1991.

Vol. 1478: M.-P. Malliavin (Ed.), Topics in Invariant Theory. Seminar 1989-1990. VI, 272 pages. 1991.

Vol. 1479: S. Bloch, I. Dolgachev, W. Fulton (Eds.), Algebraic Geometry. Proceedings, 1989. VII, 300 pages. 1991.

Vol. 1480: F. Dumortier, R. Roussarie, J. Sotomayor, H. Żoładek, Bifurcations of Planar Vector Fields: Nilpotent Singularities and Abelian Integrals. VIII, 226 pages. 1991.

Vol. 1481: D. Ferus, U. Pinkall, U. Simon, B. Wegner (Eds.), Global Differential Geometry and Global Analysis. Proceedings, 1991. VIII, 283 pages. 1991.

Vol. 1482: J. Chabrowski, The Dirichlet Problem with L^2-Boundary Data for Elliptic Linear Equations. VI, 173 pages. 1991.

Vol. 1483: E. Reithmeier, Periodic Solutions of Nonlinear Dynamical Systems. VI, 171 pages. 1991.

Vol. 1484: H. Delfs, Homology of Locally Semialgebraic Spaces. IX, 136 pages. 1991.

Vol. 1485: J. Azéma, P. A. Meyer, M. Yor (Eds.), Séminaire de Probabilités XXV. VIII, 440 pages. 1991.

Vol. 1486: L. Arnold, H. Crauel, J.-P. Eckmann (Eds.), Lyapunov Exponents. Proceedings, 1990. VIII, 365 pages. 1991.

Vol. 1487: E. Freitag, Singular Modular Forms and Theta Relations. VI, 172 pages. 1991.

Vol. 1488: A. Carboni, M. C. Pedicchio, G. Rosolini (Eds.), Category Theory. Proceedings, 1990. VII, 494 pages. 1991.

Vol. 1489: A. Mielke, Hamiltonian and Lagrangian Flows on Center Manifolds. X, 140 pages. 1991.

Vol. 1490: K. Metsch, Linear Spaces with Few Lines. XIII, 196 pages. 1991.

Vol. 1491: E. Lluis-Puebla, J.-L. Loday, H. Gillet, C. Soulé, V. Snaith, Higher Algebraic K-Theory: an overview. IX, 164 pages. 1992.

Vol. 1492: K. R. Wicks, Fractals and Hyperspaces. VIII, 168 pages. 1991.

Vol. 1493: E. Benoît (Ed.), Dynamic Bifurcations. Proceedings, Luminy 1990. VII, 219 pages. 1991.

Vol. 1494: M.-T. Cheng, X.-W. Zhou, D.-G. Deng (Eds.), Harmonic Analysis. Proceedings, 1988. IX, 226 pages. 1991.

Vol. 1495: J. M. Bony, G. Grubb, L. Hörmander, H. Komatsu, J. Sjöstrand, Microlocal Analysis and Applications. Montecatini Terme, 1989. Editors: L. Cattabriga, L. Rodino. VII, 349 pages. 1991.

Vol. 1496: C. Foias, B. Francis, J. W. Helton, H. Kwakernaak, J. B. Pearson, H_∞-Control Theory. Como, 1990. Editors: E. Mosca, L. Pandolfi. VII, 336 pages. 1991.

Vol. 1497: G. T. Herman, A. K. Louis, F. Natterer (Eds.), Mathematical Methods in Tomography. Proceedings 1990. X, 268 pages. 1991.

Vol. 1498: R. Lang, Spectral Theory of Random Schrödinger Operators. X, 125 pages. 1991.

Vol. 1499: K. Taira, Boundary Value Problems and Markov Processes. IX, 132 pages. 1991.

Vol. 1500: J.-P. Serre, Lie Algebras and Lie Groups. VII, 168 pages. 1992.

Vol. 1501: A. De Masi, E. Presutti, Mathematical Methods for Hydrodynamic Limits. IX, 196 pages. 1991.

Vol. 1502: C. Simpson, Asymptotic Behavior of Monodromy. V, 139 pages. 1991.

Vol. 1503: S. Shokranian, The Selberg-Arthur Trace Formula (Lectures by J. Arthur). VII, 97 pages. 1991.

Vol. 1504: J. Cheeger, M. Gromov, C. Okonek, P. Pansu, Geometric Topology: Recent Developments. Editors: P. de Bartolomeis, F. Tricerri. VII, 197 pages. 1991.

Vol. 1505: K. Kajitani, T. Nishitani, The Hyperbolic Cauchy Problem. VII, 168 pages. 1991.

Vol. 1506: A. Buium, Differential Algebraic Groups of Finite Dimension. XV, 145 pages. 1992.

Vol. 1507: K. Hulek, T. Peternell, M. Schneider, F.-O. Schreyer (Eds.), Complex Algebraic Varieties. Proceedings, 1990. VII, 179 pages. 1992.

Vol. 1508: M. Vuorinen (Ed.), Quasiconformal Space Mappings. A Collection of Surveys 1960-1990. IX, 148 pages. 1992.

Vol. 1509: J. Aguadé, M. Castellet, F. R. Cohen (Eds.), Algebraic Topology - Homotopy and Group Cohomology. Proceedings, 1990. X, 330 pages. 1992.

Vol. 1510: P. P. Kulish (Ed.), Quantum Groups. Proceedings, 1990. XII, 398 pages. 1992.

Vol. 1511: B. S. Yadav, D. Singh (Eds.), Functional Analysis and Operator Theory. Proceedings, 1990. VIII, 223 pages. 1992.

Vol. 1512: L. M. Adleman, M.-D. A. Huang, Primality Testing and Abelian Varieties Over Finite Fields. VII, 142 pages. 1992.

Vol. 1513: L. S. Block, W. A. Coppel, Dynamics in One Dimension. VIII, 249 pages. 1992.

Vol. 1514: U. Krengel, K. Richter, V. Warstat (Eds.), Ergodic Theory and Related Topics III, Proceedings, 1990. VIII, 236 pages. 1992.

Vol. 1515: E. Ballico, F. Catanese, C. Ciliberto (Eds.), Classification of Irregular Varieties. Proceedings, 1990. VII, 149 pages. 1992.

Vol. 1516: R. A. Lorentz, Multivariate Birkhoff Interpolation. IX, 192 pages. 1992.

Vol. 1517: K. Keimel, W. Roth, Ordered Cones and Approximation. VI, 134 pages. 1992.

Vol. 1518: H. Stichtenoth, M. A. Tsfasman (Eds.), Coding Theory and Algebraic Geometry. Proceedings, 1991. VIII, 223 pages. 1992.

Vol. 1519: M. W. Short, The Primitive Soluble Permutation Groups of Degree less than 256. IX, 145 pages. 1992.

Vol. 1520: Yu. G. Borisovich, Yu. E. Gliklikh (Eds.), Global Analysis – Studies and Applications V. VII, 284 pages. 1992.

Vol. 1521: S. Busenberg, B. Forte, H. K. Kuiken, Mathematical Modelling of Industrial Process. Bari, 1990. Editors: V. Capasso, A. Fasano. VII, 162 pages. 1992.

Vol. 1522: J.-M. Delort, F. B. I. Transformation. VII, 101 pages. 1992.

Vol. 1523: W. Xue, Rings with Morita Duality. X, 168 pages. 1992.

Vol. 1524: M. Coste, L. Mahé, M.-F. Roy (Eds.), Real Algebraic Geometry. Proceedings, 1991. VIII, 418 pages. 1992.

Vol. 1525: C. Casacuberta, M. Castellet (Eds.), Mathematical Research Today and Tomorrow. VII, 112 pages. 1992.

Vol. 1526: J. Azéma, P. A. Meyer, M. Yor (Eds.), Séminaire de Probabilités XXVI. X, 633 pages. 1992.

Vol. 1527: M. I. Freidlin, J.-F. Le Gall, Ecole d'Eté de Probabilités de Saint-Flour XX – 1990. Editor: P. L. Hennequin. VIII, 244 pages. 1992.

Vol. 1528: G. Isac, Complementarity Problems. VI, 297 pages. 1992.

Vol. 1529: J. van Neerven, The Adjoint of a Semigroup of Linear Operators. X, 195 pages. 1992.

Vol. 1530: J. G. Heywood, K. Masuda, R. Rautmann, S. A. Solonnikov (Eds.), The Navier-Stokes Equations II – Theory and Numerical Methods. IX, 322 pages. 1992.

Vol. 1531: M. Stoer, Design of Survivable Networks. IV, 206 pages. 1992.

Vol. 1532: J. F. Colombeau, Multiplication of Distributions. X, 184 pages. 1992.

Vol. 1533: P. Jipsen, H. Rose, Varieties of Lattices. X, 162 pages. 1992.

Vol. 1534: C. Greither, Cyclic Galois Extensions of Commutative Rings. X, 145 pages. 1992.

Vol. 1535: A. B. Evans, Orthomorphism Graphs of Groups. VIII, 114 pages. 1992.

Vol. 1536: M. K. Kwong, A. Zettl, Norm Inequalities for Derivatives and Differences. VII, 150 pages. 1992.

Vol. 1537: P. Fitzpatrick, M. Martelli, J. Mawhin, R. Nussbaum, Topological Methods for Ordinary Differential Equations. Montecatini Terme, 1991. Editors: M. Furi, P. Zecca. VII, 218 pages. 1993.

Vol. 1538: P.-A. Meyer, Quantum Probability for Probabilitists. X, 287 pages. 1993.

Vol. 1539: M. Coornaert, A. Papadopoulos, Symbolic Dynamics and Hyperbolic Groups. VIII, 138 pages. 1993.

Vol. 1540: H. Komatsu (Ed.), Functional Analysis and Related Topics, 1991. Proceedings. XXI, 413 pages. 1993.

Vol. 1542: J.Fröhlich, Th.Kerler, Quantum Groups, Quantum Categories and Quantum Field Theory. VII, 431 pages. 1993.

Vol. 1544: M.Schürmann, White Noise on Bialgebras. VII, 146 pages. 1993.

General Remarks

Lecture Notes are printed by photo-offset from the master-copy delivered in camera-ready form by the authors of monographs, resp. editors of proceedings volumes. For this purpose Springer-Verlag provides technical instructions for the preparation of manuscripts. Volume editors are requested to distribute these to all contributing authors of proceedings volumes. Some homogeneity in the presentation of the contributions in a multi-author volume is desirable.

Careful preparation of manuscripts will help keep production time short and ensure a satisfactory appearance of the finished book. The actual production of a Lecture Notes volume normally takes approximately 8 weeks.

Authors of monographs receive 50 free copies of their book. Editors of proceedings volumes similarly receive 50 copies of the book and are responsible for redistributing these to authors etc. at their discretion. No reprints of individual contributions can be supplied. No royalty is paid on Lecture Notes volumes.

Authors and volume editors are entitled to purchase further copies of their own book and other Springer mathematics books for their personal use, at a discount of 33,3 % directly from Springer-Verlag.

Commitment to publish is made by letter of intent rather than by signing a formal contract. Springer-Verlag secures the copyright for each volume.

Addresses:

Professor A. Dold
Mathematisches Institut, Universität Heidelberg
Im Neuenheimer Feld 288
W-6900 Heidelberg
Federal Republic of Germany

Professor B. Eckmann
Mathematik, ETH-Zentrum
CH-8092 Zürich, Switzerland

Professor F. Takens
Mathematisch Instituut
Rijksuniversiteit Groningen
Postbus 800
NL-9700 AV Groningen
The Netherlands

Springer-Verlag, Mathematics Editorial
Tiergartenstr. 17
W-6900 Heidelberg
Federal Republic of Germany
Tel.: *49 (6221) 487-410

Lecture Notes in Mathematics

This series reports on new developments in mathematical research and teaching-quickly informally and at a high level. The type of material considered for publication includes

1. Research monographs
2. Lectures on a new field or presentations of a new angle in a classical field
3. Seminars on topics of current research
4. Reports of meetings, provided they are
   a) of exceptional interest and
   b) devoted to a single topic.

Texts which are out of print but still in demand may also be considered if they fall withi these categories.

In the case of a research monograph, or of seminar notes, the timeliness of a manuscript ma be more important than its form, which may be preliminary or tentative. A subject inde should be included.

In the case of conference proceedings, articles should not duplicate the journal literature an should be individually refereed. Proceedings volumes should appear within twelve month of the corresponding meeting.

Details of the editorial policy for monographs [resp. proceedings] can be found on the in side front cover of a current monograph [resp. proceedings] volume. Manuscripts should b submitted in camera-ready form according to Springer-Verlag's specifications: technica instructions or TEX-macros will be sent on request.